THE WORKSHOP AND THE WORLD

ALSO BY ROBERT P. CREASE

The Quantum Moment: How Planck, Bohr, Einstein, and Heisenberg Taught Us to Love Uncertainty, by Robert P. Crease and Alfred Scharff Goldhaber

World in the Balance: The Historic Quest for an Absolute System of Measurement

The Great Equations: Breakthroughs in Science from Pythagoras to Heisenberg

The Prism and the Pendulum: The Ten Most Beautiful Experiments in Science

Making Physics: A Biography of Brookhaven National Laboratory

The Play of Nature: Experimentation as Performance

The Second Creation: Makers of the Revolution in Twentieth-Century Physics, by Robert P. Crease and Charles C. Mann

Peace & War: Reminiscences of a Life on the Frontiers of Science, by Robert Serber with Robert P. Crease

J. Robert Oppenheimer: A Life, by Abraham Pais with Robert P. Crease

TRANSLATED BY ROBERT P. CREASE

What Things Do: Philosophical Reflections on Technology, Agency, and Design, by Peter-Paul Verbeek

American Philosophy of Technology: The Empirical Turn, edited by Hans Achterhuis

THE WORKSHOP AND THE WORLD

WHAT TEN THINKERS CAN TEACH US ABOUT SCIENCE AND AUTHORITY

ROBERT P. CREASE

W. W. NORTON & COMPANY
Independent Publishers Since 1923
New York | *London*

Printed in the United States of America
First Edition

For information about permission to reproduce selections from this book, write to Permissions, W. W. Norton & Company, Inc., 500 Fifth Avenue, New York, NY 10110

For information about special discounts for bulk purchases, please contact W. W. Norton Special Sales at specialsales@wwnorton.com or 800-233-4830

Manufacturing by LSC Communications, Harrisonburg
Book design by Marysarah Quinn
Production manager: Lauren Abbate

Library of Congress Cataloging-in-Publication Data

Names: Crease, Robert P., author.
Title: The workshop and the world : what ten thinkers can teach us about science and authority / Robert P. Crease.
Description: New York : W. W. Norton & Company, [2019] | Includes bibliographical references and index.
Identifiers: LCCN 2018047219 | ISBN 9780393292435 (hardcover)
Subjects: LCSH: Science—Methodology. | Science—Social aspects. | Science—Political aspects. | Science and state. | Discoveries in science.
Classification: LCC Q175 .C8858 2019 | DDC 306.4/5—dc23
LC record available at https://lccn.loc.gov/2018047219

W. W. Norton & Company, Inc., 500 Fifth Avenue, New York, N.Y. 10110
www.wwnorton.com

W. W. Norton & Company Ltd., 15 Carlisle Street, London W1D 3BS

1 2 3 4 5 6 7 8 9 0

For Stephanie,
undeniably

CONTENTS

INTRODUCTION

IN THE SUMMER OF 2018, I went to see the Mer de Glace, the longest glacier in France. I knew what it looked like—or thought I did. For nearly three centuries it has been one of the most painted, photographed, and described natural features in Europe. From the northern slopes of Mont Blanc, the highest mountain in the Alps, it twists its way slowly and inexorably between the peaks like a giant icy crocodile. Its jagged white blocks inspired Goethe, Wordsworth, and other poets. In Mary Shelley's novel *Frankenstein*, the glacier's wildness is the backdrop for the monster's first confrontation with the creator who abandoned him. Many artists, including J. M. W. Turner, Caspar David Friedrich, and John Ruskin, painted its dramatic and disordered surface in images that ran from majestic and ethereal to terrifying. Visitors compared it to a hurricane-whipped ocean that had suddenly frozen and turned sheet-white.

I boarded a rack railway that had been built in 1908 to ferry tourists from the French town of Chamonix—a ski resort and hiking center near France's border with Italy—to Montanvert, a spot in the mountains near where they could step onto the glacier. The trip took twenty minutes. I found myself amid pine trees on one side of a relatively straight canyon lined by two rock walls. The ground was mossy, with no trace of snow or

ice, and the mighty glacier was not in sight. To see it, I was told, I either had to hike down or take a cable car. I hiked.

The trail ambled through trees and bushes. After a minute or so, I came to a sign in a patch of purple foxglove flowers: LEVEL OF THE GLACIER, 1820.

A few minutes later, farther down, I came across a similar marker on a lichen-speckled granite boulder. This one had the date 1890. Still later, after descending concrete treads, I passed markers reading 1920 and then 1985. Though these spots were once high points of the glacier, I saw no trace of ice or snow. The concrete treads changed to moveable aluminum stairs, the canyon walls grew steeper, and the next markers were not attached to boulders but directly to the valley's sheer rock walls. Forty-nine steps down took me to 1990, 86 more to 2001, 14 more to 2003, and 148 more to 2010. Still no ice. I was beginning to have the morbid feeling that I was descending into a large coffin. After 61 more steps I came to the final marker: 2015.

At this spot, just three years before, I would have been standing on a cold, white block of ice; instead I stared at dry rock. The glacier had melted so much that its tip had receded farther up into the valley, and I had to walk a few hundred more feet before I could get to it. Overall, my hike had taken me some 2,000 feet down vertically and about one-third of a mile horizontally. I was now at the bottom of a valley, about one-half mile across, which the Mer de Glace had once filled. It was drab and rocky, with no vegetation; occasionally ribbon-like wisps of clouds blew by overhead. Not only was the once-mighty glacier miniscule compared to its past portraits, but also its surface was flat and gray, covered with rocks and dirt.

The Mer de Glace is melting. How quickly, and what would become of it? I had no idea.

At the glacier I met Luc Moreau, a glaciologist associated with the EDYTEM Laboratory, cosponsored by the Savoie Mont Blanc University and the National Center for Scientific Research of France.[1] Tall and brawny, Moreau's robust physique is work-related. His web page displays pictures of him zip-lining over ravines and straddling crevasses while tracking the structure and movement of glaciers. He works at the Mer de Glace, measuring and helping to model it. A few weeks before, he had installed the final marker that I had seen.

Of all the Earth's surface features, Moreau told me, glaciers are the most reactive to climate. "Climate makes them, and climate takes them away." The Mer de Glace fluctuates annually, accumulating snow and ice in winters, which it sheds in summers. But overall it is not only melting but slowing down—and as the signs show, at an alarming rate. In the nineteenth century, a short path was enough to take visitors from the trailhead where I had begun. For a brief period in the 1970s the glacier swelled, but soon continued to recede. By the 1980s, the trail had grown so long that a cable car was built to ferry less venturous travelers to the glacier. In the 2000s, the tip of the glacier receded past the cable-car stop, and aluminum stairs and ramps were added. The melting is continuing at an ever-faster pace. In the 21 years between 1995 and 2016, its height dropped the same amount as in the previous 170 years. The glacier's melting has made its surface unsafe for visitors, and Moreau and other glaciologists dig an ice cave each spring that visitors can enter, covering the ice above the cave with white tarps to keep it cold and stable.

Moreau and I watched a cable car unload a new cluster of visitors. Each year, the glacier's continued melt means the ice cave has to be recut in a new spot, farther from the cable-car stop, and the ramps to it from the cable-car stop extended. Eventually, adding ramps will not make sense, and plans are being drawn up for a new cable car. I asked Moreau how much longer such fixes could go on. "I don't know," he said.

We were standing at the bottom of the half-mile-wide valley. We could even hear the glacier melt, as droplets falling from the cave roof splattered on the floor below. Outside, the desolate basin was mostly quiet and still. But every few minutes, we heard a strange creaking and cracking in the distance as a boulder broke loose and tumbled down the valley walls, knocking into and dislodging other rocks, each leaving behind a comet-trail of dust until everything came to rest on the glacier. It was a disturbing experience. It felt like the world was falling apart.

MELTING GLACIER

The Mer de Glace is melting. How fast? Glaciologists study this question with data from a variety of equipment including tools to extract and analyze ice cores, instruments to monitor ice flow and strain, and Earth- and satellite-based information collection systems. They also rely on data from other scientists: chemists, physicists, engineers, and climatologists. Integrating this information and applying mathematical methods, the glaciologists create models of what the glacier looked like in the past, its current behavior, and future prospects, and are constantly revising these models as new data comes in. When Moreau told me the results, it was therefore not his opinion. Rather, it was a picture that has been painstakingly produced and evaluated by a coordinated ongoing interdisciplinary network of scientists in what I'll refer to broadly as the scientific "workshop."

According to this picture, glaciers have grown and shrunk in response to changes in the Earth's climate over the past two-and-one-half million years. These changes are mainly due to changes in the composition of the Earth's atmosphere and the way it absorbs heat from sunlight. Four centuries ago, the Mer de Glace was about the height it is now. Then, between about 1300 and 1850, the Northern Hemisphere underwent what glaciologists call a "Little Ice Age." Several events, including changes in oceanic and atmospheric circulations and a large number of volcanic eruptions, made the Earth's average overall temperature fall by about 1°C. This small amount had huge climatic effects, making the Mer de Glace swell until it filled the valley.

Starting around the nineteenth century, however, the Earth's climate changed—first slowly, then more speedily—and for disturbing reasons. Fossil fuel burning increased the amount of carbon dioxide in the atmosphere. In the middle of the Little Ice Age, about 1620, the level of carbon dioxide was about 270 parts per million (ppm). In the last century, its concentration climbed steeply, and has now passed 410 ppm. A key "greenhouse gas," carbon dioxide absorbs infrared radiation from the Earth's surface, which the nitrogen and oxygen that compose 99% of the Earth's atmosphere cannot. The carbon dioxide molecules then transfer this energy to the nitrogen and oxygen molecules, warming the atmosphere overall. Rising concentrations of carbon dioxide—by the amount one would expect from consumption of fossil fuels—is the main reason why Earth's average atmospheric temperature has increased by 1°C in the past century alone. Again, this temperature rise seems small—yet it has caused major changes on the Earth's surface, including melting polar ice, rising sea levels, ocean acidification, vanishing coral reefs, vanishing of organisms that depend on those reefs, and migrations and extinctions of species.[2]

And, of course, glacial melting.

The Mer de Glace is concrete evidence of global warming. Each year, Moreau said, it loses about a dozen more feet of ice height. Because humanity appears unlikely to stop pumping carbon dioxide into the air, the Earth's temperature will continue to rise, by an estimated 2°–6°C for the rest of this century. The Mer de Glace and other glaciers will go on melting; many will vanish entirely. This is bad news, since glaciers account for a large fraction of the fresh water used by the Earth's inhabitants.

MELTING AUTHORITY

As Moreau and I spoke, we were aware of an alarming reality: many US politicians react to this picture by accusing scientists of dishonesty, false confidence, and being out of touch. That reaction—of rejecting the authority of the workshop—is known as science denial, and it is now an established feature of the US political landscape. A loaded and politicized term, science denial refers not to the outright rejection of scientific authority—for its practitioners still consult doctors about their health, weather.com about today's temperature, and engineers about the safety of buildings—but only in certain areas where political, economic, and religious interests come into play. In these areas, certain American politicians have found that the conclusions of the scientific workshop obstruct their goals rather than help to realize them, and therefore treat it not as an aid to best practices but as a political opponent. They lodge several different kinds of accusations to defend their rejection. Some claim that global warming is a hoax perpetrated by scientists with hidden agendas.[3] An array of other politicians have said that "I am not a scientist" and do not need to be, for science is an abstract practice of little relevance to the concrete world of politics.[4] Still others point out that science is uncertain and the jury is still out for the complex models used to predict global warming.[5]

Though damaging to the safety and welfare of citizens, and to the institutions that protect them, science denial is difficult to deter. The reason why is astonishing—its practitioners are exploiting real vulnerabilities in science itself. Take the three accusations I mentioned, that science can be used to promote hidden agendas, is abstract, and is uncertain. These are not wholly implausible. First, let me point out that a scientific workshop is a collective—a bureaucracy—whose internal politics can shape how a result is presented.[6] Second, understanding raw data and acquiring the expertise to transform this data into a meaningful finding about the world involves a level of training and expertise out of the reach of the average citizen and politician.[7] Third, science is innately uncertain, perpetually open to revision on the basis of new information.[8] These three aspects of science—and a few more I'll discuss in this book—fuel science denial. Money and political stakes play a huge role, but they only exploit core structures of science itself. Without these features, science denial would be implausible, regardless of how much funding or political power is at stake. When one does not take these features into account, attempts to stop science denial are doomed to an endless game of political whack-a-mole; instances of science denial will simply spring up someplace else. Attacks on science denial that deny the existence of these aspects—that insist that scientists must be obeyed, and when they aren't, attribute it to ignorance, irrationality, or mendacity—are as fraudulent as science denial itself. Such attacks are dangerous, for they are yet another way of misunderstanding the kind of authority that science has.

Some people, including many scientists, seem resigned to this. They hope that scientific authority is a natural thing that will shortly reassert itself, like a sturdy self-righting boat knocked over by a rogue wave. The features I mention above guarantee that such a tranquil recovery will not happen. Science in this respect might be compared with Facebook. The very gears that make Facebook socially wonderful—its ease of connecting and sharing—are the same ones that facilitate trolling, the flourishing of hate groups, the dissemination of fake news, and dirty political tricks. In a similar way, the gears that make science work—the fact that it is done by collectives, is abstract, and always open to revision—also provide fuel for science deniers. Given that this makes naïve hope unrealistic, another response might be to become filled with rage at the dishonesty and naked self-interest of science deniers in view of the inevitable damage that their efforts will bring to human life and the environment.

The chapters that follow will explain how the current state of affairs came about, and what will be necessary to change it. Aristotle, one of the most practical and wise of all philosophers, wrote that, while it is easy to become angry, it is harder to be angry "with the right person, and to the right degree, and at the right time, and for the right purpose, and in the right way." This book is about how to get angry about science denial in the right way.

REVERSING THE MELTDOWNS

Part One of this book is about the first articulation of scientific authority, told through the stories of Francis Bacon (1561–1626), Galileo Galilei (1564–1642), and René Descartes (1596–1650). They were born in an era that knew two principal sources of authority: spiritual and secular. The Church claimed spiritual authority, while the government claimed secular authority, meaning on any nonspiritual matters. Bacon, Galileo, and Descartes were among the first to describe a third kind of authority: scientific authority. This new kind of authority was grounded in the structure of Creation itself. Bacon laid out an ambitious vision of the scientific workshop that would be needed to discover this structure. Galileo defended the authority of science, arguing that its authority was as divinely grounded as that of the Church. Descartes described the special mental training—expertise, we would say—needed in the scientific workshop, arguing that such training involves sequestering oneself from, but not repudiating, the rest of the world. But as we will see, several vulnerabilities of the authority of the scientific workshop have already appeared.

Even more vulnerabilities emerge in Part Two of this book, when it becomes apparent that the workshop's findings can be oversold, treacherous, and corrosive—and scientific authority is not enough by itself to help humanity ward off its threats and realize its hopes. This part is told by the stories of such thinkers as Giambattista Vico (1668–1744), Mary Shelley (1797–1851), and Auguste Comte (1798–1857). Pursuing the scientific method single-mindedly outside the workshop is toxic to human cultural life, Vico argued, and if taught to the exclusion of the humanities, it makes people "go mad rationally." Mary Shelley's novel *Frankenstein* rang alarm bells for its still-current warning that single-minded pursuit of scientific goals is not always liberating, and the vast and at times incomprehensible power of human interactions with nature creates a potential for tragedy. Auguste Comte realized that science alone will not protect us from natural hazards or bring about social peace, and the relationship between the workshop and the world needs to be specially cultivated. These thinkers therefore saw the authority of science as having less to do with its tie to Creation than to the way it is practiced.

Part Three is about sophisticated attempts to understand the deeply complicated relationship between the workshop and the world, told through the stories of Max Weber (1864–1920), Kemal Atatürk (1881–1938) and his precursors, and Edmund Husserl (1859–1938). Weber, one of the great scholars of the subject, saw that all forms of authority, including that of science, arise from a complicated mixture of ever-changing factors. He also foresaw problems in the inevitable bureaucratization of science, which can appear to introduce lack of concern for human values. Atatürk, and his predecessors who founded modern Turkey, realized that, in the end, the authority of the scientific workshop and its findings rests on people, not tools or methods. Its authority depends on how we answer the questions, "Who are we?," "Whose science and technology is it?," and "What will it do to us?" Husserl realized that some workshop findings can dazzle human beings, making the surrounding world—the one in which we make friendships, play and work, breathe and suffer, hope and fear—appear to recede in importance, "subjective" compared to the "objective" output of workshops. This fading in importance of the humanities, he thought, was behind the cultural crisis of his time, most notably in the form of Nazism.

Part Four, the final part of this book, is about reinventing authority.

Here we examine the work of a single person, Hannah Arendt (1906–1975). A passionate and perceptive thinker who barely escaped the Holocaust and lived through a space and time when spiritual authority entirely disappeared, she provided a deep analysis of authority, as well as clues for how to restore it. Authority flourishes not in facts but in the institutions that produce and preserve them. Her work, which represents the practice of the humanities at its best, points beyond the disbelief, diatribe, and easy moralism with which science denial is usually treated.

These ten individuals each confronted severe problems with scientific authority in their times, reacted with different forms of anger, and took action. Some risked their lives. Taken together, their stories show why the dwindling authority of science is as threatening to human life as the dwindling glacier, and what can be done to keep our world from falling apart. Their stories can help show us how to respond in the right way, at the right time, and toward the right people, with regard to those markers on the trail to the Mer de Glace—and anywhere else that science denial threatens public health, the welfare of future generations, and the fate of the planet.

THE WORKSHOP AND THE WORLD

I

ONE

FRANCIS BACON'S NEW ATLANTIS

IN 1624, an English vessel set sail from Peru westbound across the Pacific with a year's worth of supplies. For a few months, the winds blew favorably; but then they changed direction, stalling the boat's progress and blowing the sailors far off course. Struggling against the elements, they exhausted their provisions, and many fell sick. Utterly lost and abandoned to nature, they gave up hope, prayed, and prepared to die.

A miracle saved them. Cloud formations appeared that typically indicate land, and the sailors headed that way. Soon they came upon an island that was not on their charts but had a small, well-built port. After landing and disembarking, the crew discovered that the community, called Bensalem, was ruled by an academy of scholars called Salomon's House. Its leaders were priests, and they used science and technology to improve their lives. The sailors were cured, taught about the island, and freed to tell the world what they had discovered.

Is this science fiction? No. It's a parable called *New Atlantis*, written about 1624 before the dawn of modern science, by the philosopher and politician Francis Bacon (1561–1626). Like all parables, it is easy to decode and digest. The vessel represents humanity making its long

Francis Bacon (1561-1626).

journey in an unpredictable and threatening world. Without special knowledge, humans get weary, sick, and lost. When finally saved, their instinct is to call it a miracle. This is an illusion. Humans can save themselves only by understanding and controlling nature. That is best done, Bacon thought, by creating an extensive and integrated scientific infrastructure that plans and executes its investigations in a systematic and coordinated way.

Atlantis was a doomed fictional island-state described by Plato to portray how political empires malfunction. In the *New Atlantis*, Bacon appropriated that tale to pass on his vision of how modern political states can succeed. Success, he thought, requires harmony between a well-organized scientific workshop and a world prepared to support and benefit from it. He expressed his vision in many ways—through whimsy, imagery, arguments, blueprints, and parables—but he did it so passionately and aggressively that he began to collect a growing list of powerful enemies who would ultimately bring him down. To a large extent, the modern world embraces his vision. The United States alone, for instance, has established dozens of Federally Funded Research and Development Centers that conduct research for the US government, covering fields from medicine to cosmology. These and other university and privately supported laboratories constitute the modern scientific workshop, our version of Salomon's House. But examining Bacon's original vision will provide a beginning for understanding science denial—that is, resistance to this workshop.

WHIMSY

Though Bacon was born to a wealthy family who lived in London, his world was full of perils and insecurity. Epidemics periodically swept through the city, including one in 1563, two years after Bacon's birth, that decimated the population. Medical care mostly took place at home with no antibiotics or anesthesia, and no awareness of bacteria. Children suffered disproportionately; they could be put to work when only eight years old, and were especially vulnerable to accidents and illness. Streets lacked drains and sewers, teemed with rats and garbage, and were dark and dangerous by night. London's elites preferred to live along the Thames River so they could travel by boat when possible.

Bacon was born at York House, one of several imposing mansions that stood between the Strand, a prominent street running along the Thames, and the river itself. Today the only vestige of York House is a small

The Watergate, dock of York House, Bacon's birthplace.

Roman temple-like structure known as the Watergate; it serves as the well-lit backdrop for an open-air bar in Victoria Embankment Gardens in downtown London. The Watergate was installed the last year of Bacon's life, so it was not the dock that he knew. Still, the fact that mansions of the wealthy frequently had docks so they could avoid street travel is a reminder of the conditions that Bacon sought to improve.[1]

Bacon grew up at the center of political power. His father Nicholas was a trusted adviser to Queen Elizabeth whose title was "Keeper of the Seal," and his mother Anne was an educator and translator. While York House was their city dwelling, their country home was Gorhambury, a mansion that Nicholas built thirty miles north of London, about a four-to-five-hour carriage ride from York House.[2] Over the fireplace at Gorhambury hung a painting showing Ceres, the goddess of agriculture, teaching human beings to sow grain. Each day at meals, the goddess would gaze down at young Francis above the words *Moniti Meliora*: "instruction brings progress."

Nicholas would often bring his precocious child Francis to court, where the charmed Queen nicknamed him "Young Lord Keeper." Like his father, Francis attended Cambridge University, entering at age twelve. Francis spent three years at Cambridge, except for a few months when it closed during another epidemic. Wanting Francis to have a political career, Nicholas pushed Francis toward law, then as now the standard path into politics. Graduating from Cambridge in 1576, Francis followed his father's footsteps and enrolled at Gray's Inn, the professional association for barristers in London. Nicholas also arranged for his son to gain diplomatic experience by joining the entourage of the ambassador to the French court.

In 1579, these ambitions screeched to a halt when Nicholas died unexpectedly of natural causes. He had arranged an inheritance for his two sons from a first marriage and for Bacon's elder brother Anthony, but had not completed arrangements for Francis, who returned from France for the funeral suddenly jobless and poor. Francis, now eighteen, had to move to quarters at Gray's Inn and start practicing law to earn money rather than as a political steppingstone.

Francis took well to life at Gray's Inn. A compound that included a law school, a dormitory, a chapel, and a series of offices, it nurtured young lawyers. Like a college campus, it teemed with social rituals and lively parties. Francis held positions in its administration and wrote speeches and plays to be performed at holidays.[3] In one scene of a play that was performed during the Christmas holiday of 1594–1595, he whimsically envisioned a government adviser, much like what he wanted to become, pleading to a ruler, much like Elizabeth, to embark on a project to seek out, invent, and discover all the secrets of nature. As part of this program, the ruler would set up a library of books from all times and in all languages, a zoo whose occupants included the full range of animal and vegetable specimens, and laboratories and museums containing all known inventions and instruments. The result, according to the play, would be to elevate the ruler who enacted this program to the status of "the only miracle and wonder of the world."[4]

ARGUMENT

At Gray's Inn, Bacon tried to jump-start his political career on his own. This was tricky and often dangerous in Elizabethan England. The Queen surrounded herself with hundreds of noblemen and advisers whose influence depended on their ability to serve, flatter, and please. Most government officials, including members of Parliament, were unpaid, and were either wealthy noblemen or were otherwise supported by noblemen who wanted influence. These officials could fall in and out of favor quickly, losing their roles and sometimes lives in the process. One had to be flexible and persistent to survive that environment.

Bacon leapt right in. When still another epidemic killed a distant relative who was a member of Parliament, he used patronage and family connections to become the replacement. Thanks to his acquaintance with the Queen, he also managed to become her informal legal adviser. As a political mentor, he selected the handsome and charismatic Earl of Essex, a politically ambitious general and confidant of the Queen.

This turned out not to be a wise choice. The ambitious Essex soon became a political rival of the Queen. In February 1601, he marched through London with soldiers trying to rouse enough people to overthrow the monarchy. The coup failed, and Essex was arrested, tried for treason, and condemned to death. During Essex's trial, the Queen asked Bacon to help prepare the case against his former mentor. Bacon's defection from his former mentor provided fodder to a growing number of people who regarded the irrepressible Bacon as an egotistical opportunist.

Elizabeth died in March 1603—another plague delayed her funeral—and was succeeded by James I. Bacon's assiduous efforts in Parliament, and his prolific advisery memos, attracted the new King's attention. Bacon was soon knighted. He also met and courted Alice Barnham, the eleven-year-old daughter of a London merchant and politician. When they married, Francis was forty-six and Alice a few days shy of fourteen, and her dowry helped him considerably. Their marriage, plus Bacon's numerous friendships with younger men and indications of coolness in the marriage, has sometimes raised speculation that he was gay and needed a cover. But marriages to much younger women were common among members of Bacon's circle, and no evidence exists of his being a homosexual.

When the next plague swept London, Parliament went into recess. Bacon used the time to write *Advancement of Learning*, an extended argument for his program of science and education. Kings, Bacon argued, have a moral duty to promote the study and control of nature, but in order to do so they have to break with the past. We cannot keep following in Aristotle's footsteps and focus on first principles, formal proofs, and argument-making, Bacon wrote. That works for topics like geometry, but not for finding out about God's Creation. A handful of discoveries—gunpowder, the compass needle, and the printing press—had transformed the world more than any war. But these had been discovered by accident! What greater glory for a King, and benefit for his subjects, than to devote the power of his kingdom to seeking discoveries systematically?

Bacon's book was the first comprehensive articulation of the rationale for experimental science and its possibilities for improving the human condition through systematic study. He targeted the book's language at James and sent copies to James and his advisers. Its closest modern equivalent is "Science: The Endless Frontier," crafted by the US engineer and science administrator Vannevar Bush in 1945. Commissioned by President Franklin D. Roosevelt, Bush's report crystallized the lessons of the wildly successful wartime application of science and technology into a plan for their continuing development in peacetime. Bacon is not mentioned, but his influence is clear throughout the report. Its last line—"On the wisdom with which we bring science to bear against the problems of the coming years depends in large measure our future as a nation"—could have been written by Bacon himself.[5] That report shaped US science policy for decades, and provided part of the rationale for the creation of the National Science Foundation. Its influence, however, has waned, and a twenty-first-century revision updating the arguments for maintaining a diverse, extensive, and globally connected scientific infrastructure is badly needed.

Bacon published the *Advancement of Learning* in the fall of 1605. It was a great book, but the timing was bad. That fall, London was abuzz with the revelation of the Gunpowder Plot, in which disgruntled Catholics attempted to assassinate King James and blow up Parliament. Bacon's book was buried in the resulting trials and investigations. Still, his career finally took off. In 1617, James appointed Bacon to his father's old position of Lord Keeper; the one-time "Young Lord Keeper" was now the genuine item. The next year, Bacon became Lord Chancellor, the pivotal politician who mediated between Parliament and the Crown.

APPEALS

His political career established, Bacon returned to promoting his scientific vision in earnest. He had to do two things: describe the science he had in mind, and explain its social value. It was not easy when no examples of either yet existed—no laboratories or academies, no vaccines, cellphones, or examples of specific life improvements. Most of his contemporaries believed in supernatural things like witches and werewolves, bloodletting, omens from God, and comets that portended evil: How could he go about convincing them that, to *really* improve their lives, they should not just read the Bible and pray but instead should study things like sanitation, food preservation, and energy?

Bacon experimented with different kinds of appeals. He came up with several, and he framed each as strongly as possible.

One appeal was to biblical imagery. The most widely read and authoritative book, the Bible was a storehouse of powerful and compelling images, which Bacon appropriated. One of its most forceful and familiar is of the Fall of Adam and Eve, when the human race, bewitched by a serpent, passed from innocence to sin. Bacon wrote that humans experienced a *second* Fall when their minds were bewitched by the false idea that they were supposed to simply observe the world rather than use its resources for self-improvement. Humans can overcome the first Fall, to the extent that they can, by cultivating morality and religion; they can overcome the second, to the extent that they can, by cultivating science and technology.

Another appeal was to call nature a "book." God created two books for humans, Bacon said, the Bible and Nature. In the first, we read of God's will, which includes the instruction, laid out in Genesis, that we humans rule over the fish in the sea, the birds in the heavens, and the Earth and everything on it. We can find the tools to do so, however, in the second book. Humans thus have to read the book of nature, not to *understand* better the message of scripture, but to *obey it*. Calling nature a "book" was brilliant, because the image had deep roots in Christian tradition.[6] But Bacon gave the familiar image a clever twist of far-reaching importance, for his use of it directed attention to nature as a separate field of study in its own right. The first book is for those who want to focus on how best to live in the world of social, moral, and religious life. The other book is for those who want to understand nature apart from the issue of how to live well. The two-book distinction marks the birth of the gap between the "two cultures," the humanities and the sciences. Bacon was one of the first to identify that gap. He minimized the gap, viewing the two books as complementary, with an intimate connection between human moral duty and the pursuit of science. The sense of intimate connection between these two goals—that our moral behavior alone should motivate us to do science—has been lost in the twenty-first century, making the gap seem huge.

Bacon also made powerful appeals through the use of parables and fables, real or fictional stories with built-in, easy-to-grasp morals. All his contemporaries knew the story of Oedipus and the Sphinx, for instance. According to the myth, mentioned in Sophocles's play *Oedipus*, travelers to and from ancient Thebes were liable to being ambushed and killed by a Sphinx until Oedipus came along and subdued the creature by revealing its secrets. In a book called *The Wisdom of the Ancients*, Bacon asserted that that fable is actually about science. Bacon likewise reinterpreted the fable of Proteus, the Greek god of rivers who kept changing forms when the Greek king Menelaus seized him to get information. That story was really about what we need to do to understand nature, Bacon wrote. We cannot be content to passively observe a natural phenomenon act in one particular way if we want to understand it; we have to "vex" it to see its various forms. This is, in effect, an early defense of scientific experimentation, in which inquirers into nature need to see how its features perform in different circumstances. You can't just look and describe nature to understand it; you have to perturb and intervene.

All of these appeals were ways, Bacon once said, of trying to gain "a quiet entry into minds choked and overgrown."[7] That remark is disingenuous. Bacon was aggressive and unrelenting, and "finding a way to blast through layers of complacency and ignorance using any means necessary to goad people into doing the right thing" would be more like it. At one point in the *Advancement*, Bacon compared himself to the religious revolutionary Martin Luther, whose revulsion over hypocritical religious practices led him to "awake all antiquity, and to call former times to his succours to make a party against the present time: so that the ancient authors, both in divinity and in humanity, which had long time slept in libraries, began generally to be read and resolved."[8] Bacon set to work with every means he could, including religious imagery, ancient authorities, familiar fables, powerful metaphors, and any value he thought his contemporaries accepted in an effort to entice seventeenth-century readers not to be complacent with malnutrition, disease, and threats to humanity, but to do something about them. In the twenty-first century, we need some equally impassioned revolutionary inspired by Bacon's example to similarly incite people not to be complacent about melting glaciers, rising seas, and global epidemics. These pose issues for our health and welfare, and for those of our descendants, as serious as those Bacon experienced in his time.

BLUEPRINT

In 1620, Bacon published his most important work, the *Instauratio Magna*. The title is literally translated from Latin as *The Great Instauration*, but this

Frontispiece of "The Great Instauration."

is needlessly clunky; *The Great Renewal* or even *The Great Revolution* would be a closer approximation. The frontispiece by itself is at once whimsy, argument, appeal, and parable. A three-masted galleon (the ship of learning) is venturing past the pillars of Hercules (the limits of the known world) into unknown parts of the ocean (nature). Inscribed on one scroll is *Moniti Meliora*, "instruction brings progress," the legend over the fireplace in Bacon's childhood home. Another scroll bears a biblical quotation (Daniel 12:4): "Many will pass through and knowledge will be increased."

In the preface, Bacon promised to show how to reconstruct the "sciences, arts, and all human knowledge" on "proper foundations." This may not restore Earth to "its perfect and original condition," but the better we know nature, the more we can use it. "Knowledge is power" was his basic message. He aggressively pressed the case for practicing science to improve food production, housing, medicine, and navigation. He apologized that the work is unfinished. The ideas are obvious, he said, but didn't seem to have occurred to others. Fearing they would vanish at his death, he wanted to outline his project to kindle new light for the "benefit of the human race." Dedicating this work to King James, he begged the monarch to finance his ideas. "I have provided the machine," Bacon concluded, "but the stuff must be gathered from the facts of nature." The reason, he says, is that "The chain of causes cannot by any force be loosed or broken, nor can nature be commanded except by being obeyed."[9]

In *Instauratio Magna*, Bacon provided the blueprint for a vast scientific infrastructure. It would consist of a coordinated network of workshops in which teams of researchers conduct experiments and analyze results to develop and extend their knowledge and expertise about some corner of nature, capitalizing on the knowledge and expertise of others. These workshops would churn out findings relevant to urgent issues such as health and energy, as well as communication, manufacturing, and transportation. Though a shambles, *The Great Instauration* is considered one of the most ambitious and influential works ever written. The poet Abraham Cowley cited it in comparing Bacon to Moses, who led his people out of Egypt. The poet Samuel Taylor Coleridge called the separately published *Novum Organum*, the second part of the *Instauration*, "one of the three great works since the introduction of Christianity." In the intervening centuries, Western nations have largely followed Bacon's blueprint and established such an infrastructure. That infrastructure is one of the principal means by which Western civilizations understand the surrounding world, and attempt to discover and ward off threats, address needs, and realize hopes. It is a foundation of this culture.

In the midst of this work, though, Bacon's life was unraveling. The King grew annoyed at Bacon's incessant and unsolicited foreign policy advice, and assigned him to handle controversial and high-profile legal cases, including prosecuting Sir Walter Raleigh for treason. Bacon once again began to make powerful enemies in Parliament because of his high-handed ways and ostentatious behaviour. Resentment against him mounted when he used his connections to elevate himself and obtain new titles: first Baron, then Viscount.

Bacon's soaring ambitions (both scientific and political), the Essex episode, and the bribery scandal have sometimes caused people to regard him as a seventeenth-century villain. He has been called materialistic, utilitarian, power-hungry, and heartless, "the most wicked man in recorded history," and a "creeping snake" with "viper eyes." Some of his critics invoke the poet Alexander Pope's quotable couplet: "think how Bacon shin'd/The wisest, brightest, meanest of mankind." Pope's lines were cited by others, including the early nineteenth-century British politician Lord Campbell, who claimed that Bacon was "THE MEANEST OF MANKIND!!!," using all caps and triple exclamation points for emphasis. In 1978, Anthony Burgess blamed terrorism on "a Baconian faith," claiming (falsely) that Bacon held that creating a new future requires destruction of the past, while in 1985, *Time* magazine lumped him alongside US presidents Garfield and Nixon as a famous corrupt politician.[10] That sounds damning. But these criticisms—as the scholar Nieves Mathews has shown in *Francis Bacon: The History of a Character Assassination*—pluck Bacon's remarks and actions out of context.[11] The word *meanest* in Pope's couplet probably signifies "unassuming"—which is the sense in which Bacon himself generally uses the word—rather than "villainous." Similar misreadings characterize most other accusations. The evidence against Bacon's character, one could say, has been cooked.

But his contemporaries found plenty to cook with. In 1621, Bacon's rivals formally accused him of bribery and corruption. His actions were customary for the time, but a series of scandals emboldened the government's opponents to push back aggressively against abuses. They seized on the testimony of two individuals who said they had given money to Bacon while their cases were being tried in court. They had lost, so their complaint was less that Bacon had taken money than that he had not given them its worth. He prepared to defend himself but found King James unsupportive; politically, someone's head had to roll, and Bacon's was the most obvious. Seeing the writing on the wall, he abandoned his defense and asked for the King's mercy. He was fined, briefly imprisoned in the Tower of London, and stripped of the right to hold political office.

UTOPIAN VISION

Bacon spent the rest of the time out of politics, traveling back and forth between Gray's Inn and a house he had built (which no longer exists) near Gorhambury, working out his scientific visions. One of these was *New Atlantis*, the fable mentioned at the beginning of this chapter. It is the spirit of Bacon's vision and of all his appeals—whimsy, argument, appeal, blueprint—condensed into a single utopian fable. It's what a scientific community (Salomon's House) would look like, how it might interact with the broader social community (Bensalem), and how the two might harmonize.

The fable opens with European sailors getting blown off course, falling ill and giving themselves up for lost, and happening on a previously unknown island: Bensalem. The inhabitants who first contact them, noting that many of the crew are sick, warily keep a distance until an antidote for epidemics is procured. The inhabitants also refuse to let the crew ashore until ascertaining that they are Christian (more on this in a moment). The visitors are brought to a guest lodge where they are greeted by its head, a male priest (the island's leaders are all Fathers, but more on this, too, in a moment). He then tells them the history of Bensalem and Salomon's House, and calls the latter the "lantern" or "eye" of the kingdom. He also tells them that the reason the island is not on their charts is that Bensalem wanted to remain unknown lest unscrupulous outsiders raid their flourishing land. Later, the crew meets one of the Fathers of Salomon's House, also a priest. The Father explains its mission: "the knowledge of Causes, and secret motions of things; and the enlarging of the bounds of Human Empire, to the effecting of all things possible." This was remarkable: a government agency whose purpose was not to praise God or Creation, nor make a profit, nor pursue a political agenda, but to make human life flourish.

The Father then describes some activities of Salomon's House. Bacon here envisions the extensive array of tools, devices, materials, and laboratories that would be needed in a fully functioning scientific workshop. Underground labs investigate mining, cure diseases, and examine other projects. High-altitude labs study astronomy and meteorology. Thanks to these and other facilities, the scientists of Salomon's House learned to manipulate conditions in the land, sea, and air—including, surely, those that had brought the Europeans to the island in the first place. Salomon's House has Chambers of Health—research hospitals—where scientists find ways to cure disease and strengthen human bodies, and also food laboratories to improve nutrition. There are zoos, aquariums, furnaces, pharmacies, and workspaces in which to build and test instruments. Readers of Bacon's time would notice that many of the things that the Salomon's House scientists have learned how to study and control—frogs and flies, thunder and lightning, hail and floods—are called miracles in the Bible. In reality, Bacon is saying—in anticipation of species management and weather modification—they can be controlled by humans.

Bacon realized that the administration as well as the scientific infrastructure of the workshop would have to be diverse and coordinated. Salomon's House is composed of a network of differently skilled people: teachers, publishers, technicians, and explorers ("Merchants of Light") who covertly seek out the science and technology developed by other nations. Others collect books and conceive and execute experiments. In an anticipation of modern science management, Bacon envisioned the community as having its own rituals to reinforce cohesion: awards, honors for inventors, and so forth.

The harmony between the scientists of Salomon's House and the citizens of Bensalem requires also that the citizens are prepared to support it. It would not be enough for Bensalem simply to have a productive scientific workshop. Its citizens would have to *see* that the workshop is worth having, and to value the flourishing of human life. They could not, that is, be narcissistic and concerned only with themselves, but must be concerned about their fellow humans. Only then would they experience the scientific workshop not as an intrusion or threat, nor as a special interest, but as the most effective way to promote their own values, spiritual and material.

This is how to understand the references to Christianity in the *New Atlantis*, which otherwise sounds preachy and intolerant to a modern reader. Bensalem, in fact, has a Jewish population and a measure of diversity in its population. The Christianity of Bensalem (as in England after the Reformation) is a state-sponsored religion focused not on life after death but service to the living. Its Christ is not the other-worldly member of the Trinity nor the sermonizing scold of modern politicians, but rather the this-worldly gentle saint of the gospels. In Bensalem, to be Christian means to act humanly and compassionately—"full of humanity," the narrator says—as a baseline rather than selfishly. Only in a world with those values could a scientific infrastructure be sure to benefit *all* citizens.

Bensalem thus needs no revolutionary nor relentless scold to convince its citizens to appreciate science and technology. The scientists do, however, take steps to cultivate the appreciation of its citizens, through what we might call popularizing science—by making excursions outside Salomon's House and giving lectures in Bensalem's cities. Such activities would encourage "generosity and enlightenment, dignity and splendor, piety and public spirit."

All through *New Atlantis*, the crew assumes that, in order to protect the island's secrecy, they will not be allowed to leave. At the very end, in a dramatic and unexpected turn, the Father tells the crew that they are free to leave, return home, and tell the world about Bensalem.

DIGESTING BACON

In *New Atlantis*, Bacon recognized that a scientific workshop would be a self-administering, collective enterprise. But that very collective character raises the possibility of its being viewed with suspicion. The Father tells the crew that the leaders of Salomon's House have "consultations" regarding which of their inventions and discoveries to publish and which not, and which to tell Bensalem's rulers about and which to keep secret. How do they decide? In a utopia, one can probably assume that leaders who are Christian priests are self-reflective enough not to be motivated by self-interest, concealed agendas, or hidden prejudices. Modern readers, though, are all too familiar not only with abusive and corrupt Christian priests, but also with episodes—involving weapons, pollutants, racist clinical studies, science that serves political interests, and the exploitation of natural resources to serve the rich and privileged—where scientific knowledge works not to humanity's benefit but for special interests or to reinforce existing injustices and racial inequality, sometimes deliberately and sometimes not.[12]

That can be seen even in *New Atlantis* itself. The vessel that sails into Bensalem—which represents humanity—does not have any women on board. That's possibly excusable, given the terms of the fable that Bacon was setting up; a trading vessel of that sort typically would not have had women on board. Less understandable is that, while Solomon's House has women among its attendants and servants, its leaders are Fathers. More cringe-worthy and disturbing still is Bensalem's "Feast of the Family," which celebrates fathers who have lived long enough to sire thirty descendants. During the festival, the father sits on a special chair whose canopy has been decorated by his daughters. The mother? Horrifyingly, she's kept out of sight in a room off to the side behind a concealed door. How could Bacon, a farsighted prophet of the scientific and technological age, ignore 50 percent of its population? The most charitable but still nevertheless inexcusably sexist reading is that Bacon had to be aggressive in making his case, using languages that his contemporaries understood and images they would find appealing rather than distracting, and his aim was to depict the value of science, not of gender justice. But would his readers, who were acquainted with female rulers both in legend and fact, really have been that sidetracked by a description of a place in which women participated equally in social and scientific life? All this makes Bensalem a perfect illustration of the first vulnerability, mentioned in the introduction, to which science is exposed: becoming a bureaucratic institution in which the power of knowledge is developed for use, not by all citizens, but only some.

A modern-day reader, too, is likely to experience as dangerous Bacon's enthusiasm for intervening indiscriminately throughout nature to find its secrets and to control it. His critics on this point include feminists who have pointed out that Bacon's picture mirrors his age's picture of men's rule over women. The American philosopher Sandra Harding, a distinguished research professor at the University of California, Los Angeles, sees him as advocating the "marital rape" of nature—that is, "the husband as scientist forcing nature to his wishes." Bacon, she wrote, "appealed to rape metaphors to persuade his audience that experimental method is a good thing."[13] Invoking Bacon's use of the word *vex*, another American philosopher, Carolyn Merchant, of the University of California, Berkeley, charges Bacon with "treat[ing] nature as a female to be tortured through mechanical investigations."[14] Bacon's defenders include Peter Pesic of St. John's College in Santa Fe, who shows that Bacon was using the word *vex* to mean something more like "perturb" than "torture"; in context, Bacon was advising us to see how a phenomenon acts in its myriad ways before we try to say what it is.[15] Moreover, for every passage about the need to perturb nature in order to find out about it, there is also one in which Bacon writes something to the effect that we can control nature only by understanding and obeying it. Still, reading the debates among

Ruins of Gorhambury, Bacon's childhood home.

Harding, Merchant, Pesic, and other contemporary scholars makes us aware that we see dangers where Bacon does not. We see the ability of the workshop to deliver power in a way that deliberately or inadvertently reinforces inequality and injustice and promotes hidden agendas. The suspicion that this is happening, again, is one justification sometimes cited for science denial.

Bacon died from pneumonia at age 65, in 1626, in the line of scientific duty. As the English writer John Aubrey put it in a famous account—which may not be entirely true—after leaving Gray's Inn one day, Bacon wondered whether snow might be as effective at preserving meat as salt. Bacon and his physician, who was accompanying him, resolved to test it then and there.

> They alighted out of the coach, and went into a poor woman's house at the bottom of Highgate Hill, and bought a hen, and made the woman gut it, and then stuffed the body with snow,

> and my lord did help to do it himself. The snow so chilled him, that he immediately fell so extremely ill, that he could not return to his lodgings (I suppose at Gray's Inn), but went to the earl of Arundel's house at Highgate, where they put him into a good bed warmed with a pan, but it was such a damp bed that had not been laid-in about a year before, which gave him such a cold that in two or three days . . . he died of suffocation.[16]

Two weeks after Bacon died, Alice married the man rumored to be her lover.

TODAY'S RESEARCHERS do not rank Bacon high as a scientist. He constructed no theories, devised no inventions, and has no discovery named after him. He was obtuse about the science of his own day, rejecting Copernicus's idea that the Earth moved about the Sun rather than vice versa (proposed in 1543, two decades before Bacon was born), and all but ignored his contemporary Galileo. In 2015, the Nobel laureate Steven Weinberg grumbled that "It is not clear to me that anyone's scientific work was actually changed for the better by Bacon's writing."[17]

In a narrow definition of "scientific work," Weinberg is correct. But Bacon was a prophet of his time, not ours. Writing before scientific workshops existed, he argued that nations were practically, intellectually, and morally obliged to create them for their people. In the four centuries since Bacon lived, the United States and other nations have built up a huge scientific infrastructure for researching energy, health, and the environment that in many ways resembles a global version of Salomon's House. Bacon may not have improved the work of anyone in the workshop, but he did make it possible for workshops to exist in the first place. He described and legitimized the role of the workshop that now employs Weinberg.

The original Atlantis described by Plato, a fictional island in the Atlantic Ocean, sank beneath the seas, though the textual sources do not agree on whether this was due to natural disasters or divine retribution for its inhabitants' bad behavior. The New Atlantis described by Bacon, a fictional island in the Pacific Ocean, is also precarious. Whether it sinks or survives—whether we do as well—depends on the relationship between the scientific workshop and the surrounding human community.

TWO

GALILEO GALILEI AND THE AUTHORITY OF SCIENCE

St. Paul towers over you, fury in his eyes, and you'd better listen. That's the message of "The Preaching of St. Paul," a thirteen-foot-high painting created by the seventeenth-century painter Eustache Le Sueur and commissioned for the Notre Dame Cathedral in Paris, which now hangs in the Louvre. Clad in white with an orange tunic, the fiery apostle lifts his right hand as if scolding his listeners, and clutches a book of scripture in his left. Some of his audience are fearful, while others obediently bring books to burn at his feet. One man, on his hands and knees, blows on the fire to consume the heretical material more quickly. Look carefully, and you see geometrical figures on its pages.

This painting dramatizes the predicament that Galileo and others who believed in Copernicus's heliocentric world—that the Earth and other planets trace circles or ellipses around the Sun—faced in seventeenth-century Italy. They had a choice between two views of nature. In one, contained in the book in Paul's hand, nature is God's Creation; in the other, suggested by the books at his feet, it is to be understood mechanically and mathematically. Believers in the first were told they were going to heaven, while believers in the second might get burned alive.

"The Preaching of St. Paul at Ephesus," by Eustache Le Sueur.

Galileo didn't think he had to choose, and argued that his mechanical and mathematical way of understanding nature was as authoritative as the scriptures that his opponents were citing against him. While Bacon wheedled and placated opponents, Galileo punched right back, often lacing his arguments with insults, ridicule, and sarcasm. Like Bacon, he wound up in jail—but for the audacity of his work rather than for his personal shortcomings. While Bacon was trying to motivate an indifferent King to support the benefits of science to human life, Galileo feared for his life and had to defend himself against powerful theologians who would have happily executed him.

In Europe during Galileo's time, two main sources of authority governed human life. Spiritual authority, having to do with matters relating to the soul, was claimed by the Church. Its authority stemmed from the belief that Church leaders had authentic insight into moral matters due to their special relation to the divine order. Secular authority, having to do with nonreligious issues, was claimed by the state. Its authority stemmed from the acknowledgment that the role of the state's rulers is to regulate these matters, and that citizens have an obligation to obey.

Galileo argued that a third kind of authority existed—scientific authority, the authority of the workshop. Like Bacon, he called nature a book. But while Bacon promoted the book of nature image to emphasize the value of its contents, Galileo stressed the book's origins. He wanted to demonstrate that, in Le Sueur's painting—completed six years after Galileo's death as part of the Counter-Reformation effort to combat his influence—the book about to be burnt in front of St. Paul carries as much authority as the one in his hands.[1]

CRAZY UNIVERSE

Galileo's birthplace was a modest dwelling on the Via Giusti, in the middle of Pisa a few blocks from the Arno River. It's now marked with a simple plaque. Pisa was part of the Grand Duchy of Tuscany, whose principal city was Florence. Tuscany had grown politically weaker since its glory days in the Renaissance during the fifteenth and sixteenth centuries, and jostled with other nearby principalities such as the Papal States, territories ruled by the Catholic Church.

But the real threat to Galileo came from the Church's religious authority. Until shortly before Galileo's birth, its authority had extended throughout Europe, but the Protestant Reformation had caused the Church to lose authority in parts of the north. In Italy, though, the Reformation inspired the Church to regroup and reinvigorate. A series of meetings called the Council of Trent launched what is known as the Counter-Reformation. Its objectives included reasserting Church authority over the interpretation of the Bible, rejecting the claim made by the Protestant reformers that interpreting the Bible was a personal matter.

In his youth, Galileo was unaffected by these religious matters. His father Vincenzio, a musician, sent him to the University in Pisa expecting him to become a doctor. Galileo had other ideas and became fascinated by math. His interest was partly inspired by his father's own experiments with mathematics in music. Galileo noticed mathematical relationships

Galileo Galilei (1564-1642).

everywhere: in music, astronomy, and ordinary objects. In a famous episode recounted by Galileo's later disciple Viviani, one day the young medical student took time out from praying in the Pisa cathedral to time the huge chandelier as it drifted back and forth. Using his pulse as a stopwatch, Galileo realized the chandelier swung in the same time whether traversing a long or short arc. According to the old story, this is how he discovered the principle of the pendulum: its swing depends only on its length. Modern tellers of the tale often add the historical footnote that the chandelier now in the cathedral was installed a few years after the supposed event; still, the one Galileo saw obeyed the same laws. The simplicity of the pendulum principle not only made it possible to understand the motions of all swinging objects, but it also transformed the pendulum from a toy into a valuable instrument for timekeeping and other purposes. Thanks to Galileo's observation, the pendulum is one of the oldest scientific instruments still in service—older, though just barely, than the telescope, Galileo's other instrument of choice.

Almost three hundred years later, Mark Twain visited the same Pisa cathedral and described its chandelier. It was one of the few times the American writer and satirist was awestruck.

> It looked like an insignificant thing to have conferred upon the world of science and mechanics such a mighty extension

> of their dominions as it has. Pondering, in its suggestive presence, I seemed to see a crazy universe of swinging disks, the toiling children of this sedate parent. He appeared to have an intelligent expression about him of knowing that he was not a lamp at all; that he was a Pendulum; a pendulum disguised, for prodigious and inscrutable purposes of his own deep devising, and not a common pendulum either, but the old original patriarchal Pendulum—the Abraham Pendulum of the world.[2]

Galileo continued to explore that "crazy universe"—the world of the workshop—in the coming years. Its blueprint was mathematical, he thought. Knowing math might therefore help you know the world.

In 1588, Galileo applied for a mathematics professorship in Pisa. Needing to generate buzz about his mathematical proficiency, Galileo arranged to give a popular talk on mathematics at the Florentine Academy. He chose a topic guaranteed to attract a large and appreciative audience: the shape of Dante's *Inferno*.[3]

Dante, the supreme poet of the Italian language, was beloved by inhabitants of his native Florence, who obsessively scrutinized every detail of the *Inferno*. One pressing question was the overall shape of Hell itself. Though Dante's Hell was fictional, he had depicted it so realistically that some people tried to model its overall shape with the zeal of today's Star Trek fans modeling the Starship *Enterprise*. Of the two leading models, one was proposed by a Florentine, and the other by an inhabitant of Lucca, one of Florence's most hated rivals. The *Inferno* lectures were a century-old tradition, but Galileo must have laughed when he came up with his angle, knowing it would pack the house with Florentines yearning to see him make the home team triumph.[4]

The Florentine's model was conical, like a modern-day martini glass covered by a huge dome. The rival's had a wide bowl at the top and a lower and narrower stem, like a margarita glass. Analyzing Dante's descriptions, Galileo showed that the top of Hell could not have the shape of a margarita-like bowl. More daringly, Galileo used scientific reasoning to demonstrate that, while the narrower stem shape was not inconsistent with the poem, it would not work in reality for physical reasons: the Earth's pressure would crush it inward. To fully grasp the *Inferno*, Galileo was saying, you couldn't just consult Dante's words in what we might call "Dante literalism." You also had to understand mathematics and what it said about the world.

Galileo's talk was a smash. He got the job, a three-year stint at the University of Pisa.

Then, as the physicist Mark Peterson relates, the roof literally fell in. Galileo shortly realized that the Florentine's model, too, would not actually work. Galileo had made a mistake involving scaling. His mathematics showed that a dome can easily be built over a small cone, but not over something as vast as the Florentine had proposed. If Galileo's mistake were exposed, Peterson remarks, it would mean "professional death."[5] Fortunately, it never was. Galileo hid his error for years while quietly exploring ways around it. He found none; the mathematics of scaling showed a dome that big would collapse. The episode, which deeply affected Galileo, reinforced his conviction that the structure of the world contains another source of authority, one independent of what any politician or priest says.

WORLDLY MATH

Humans have used numbers for centuries, and scholars in the ancient Greek, Indian, and Islamic worlds had produced a sophisticated body of mathematical knowledge. But until Galileo's time, mathematics was regarded chiefly as a field unto itself, and when it was used in practices like carpentry and surveying it was seen to apply to the world from the outside rather than as characterizing its structure from within. For example, when theologians attacked Copernicus for creating a model in which the Earth moved around the Sun, his followers responded persuasively by arguing that its mathematics was just a tool to predict the positions of heavenly bodies. How absurd to think that the model represented how these bodies *actually* moved!

But Renaissance scholars had rediscovered what the ancient Greeks knew, that mathematics indeed describes the world from the inside. Abraham Pendulum had suggested this mathematization to Galileo, who confronted it more directly in his *Inferno* lectures. Math was not just a tool to describe visions of the world, but rather its very syntax. A parabola was not just a convenient way to represent the motion of a cannonball, but the cannonball's actual trajectory. Those who understood mathematics and how it applied to nature, in short, were authorities of a new and different kind.

In his three years in Pisa, Galileo began to explore this new world in experiments. In 1592, he got a new job at the University of Padua, where he would spend eighteen years. There he grew into a model professor and public figure. He gave demonstrations and lectures in a hall that seated two thousand. He started an instrument-making business. He acquired followers, including a Benedictine monk named Benedetto Castelli. Handsome and outgoing, Galileo also acquired a lover, the stunning Marina Gamba. Gamba moved from Venice to Padua in 1600 to be with him when she was twenty-two years old and he was thirty-six. Galileo never married her (or anyone else) but they had three children: Virginia in 1600, Livia in 1601, and Vincenzio in 1606. By all accounts, he was a caring and supportive father.

Galileo settled in to the comfortable atmosphere and began to explore the "crazy universe," the world of the workshop, in earnest, seeking to map it mathematically. Because objects fell too quickly for him to measure, he set out to "dilute" gravity by rolling balls down various slopes. This slowed them enough so he could measure how they pick up speed. As he made the slopes progressively steeper, he could estimate how quickly the balls would fall if dropped vertically. By 1604, Galileo had discovered the mathematical law of acceleration.

He seethed, though, at the university's philosophers. It wasn't just their higher salaries and status. Mostly it was the fact that these supposed dispensers of wisdom were clueless as to how the world worked; they were usually ignorant of mathematics and proud of it. In 1604, Galileo published a pamphlet ridiculing philosophers who could not figure out whether a new star that had appeared the previous year was near the Earth, as Aristotelians insisted, or at a remote distance like the other stars, as the new mathematical approach maintained. It was easy to find the answer if you understood the mathematics of parallax, or how, from the perspective of a moving observer, an object appears to move against the background. The philosophers didn't, but pontificated about it anyway.

In 1609, a friend told Galileo about a Dutch lens grinder who had applied for a patent on a device to make distant objects seem closer. Intrigued, Galileo found he could make such a device himself, and set out to improve it. Thus began "the most intimate change in outlook which the human race had yet encountered," wrote the mathematician and philosopher Alfred Whitehead in 1925. "Since a babe was born in a manger, it may be doubted whether so great a thing has happened with so little stir."[6] Yet all the thing did was to make faraway objects seem closer.

The greatness began to stir when Galileo took the device, soon called a "telescope," outside into his garden and pointed it up. "I render infinite thanks to God for being so kind as to make me alone the first observer of marvels kept hidden in obscurity for all previous centuries," Galileo wrote shortly afterward.[7] It was not powerful, magnifying about eight times, roughly the power of a cheap set of opera glasses, but it showed the heavens to be startlingly different from how scholars had described

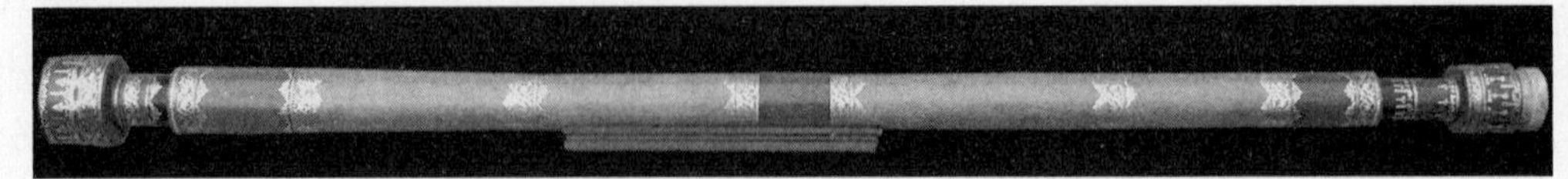

Replica of the earliest surviving telescope attributed to Galileo, on display at the Griffith Observatory.

them on the basis of observations made with the naked eye. Closer up, the Moon was full of craggy surfaces and mountains just like the Earth. The milky part of the sky was really countless numbers of individual stars. The Sun was traversed by occasional spots. Still more astonishingly, the planet Jupiter had bodies that revolved around it rather than the Earth. Galileo called these celestial bodies stars or planets rather than moons, for the concept of "*a* moon" as opposed to "*the* Moon" did not yet exist. Their discovery was perhaps the most exhilarating moment of Galileo's entire life.

In March 1610, Galileo published *Sidereus Nuncius* or "Starry Messenger." His first scholarly book, it was the first scientific work of astronomy based on telescopic observations. A few theologians were skeptical, including a colleague named Cremonini who refused to look through the telescope. But Galileo relished this opportunity to ridicule one of the earliest science deniers. "My dear Kepler," Galileo wrote the German mathematician and astronomer, "I wish that we might laugh at the remarkable stupidity of the common herd. What do you have to say about the principal philosophers of this academy who are filled with the stubbornness of an asp and do not want to look at either the planets, the moon or the telescope, even though I have freely and deliberately offered them the opportunity a thousand times? Truly, just as the asp stops its ears [as per an old legend], so do these philosophers shut their eyes to the light of truth."[8]

The telescope energized Galileo's career, giving him an innovative way to glorify his patron. He added telescopes to his instrument-making business. He dedicated *Sidereus Nuncius* to the new Grand Duke of Tuscany, his former student and friend Cosimo II de' Medici, and named Jupiter's satellites the "Medicean Stars." After the book was published, Galileo took his telescope to Pisa so he could personally show Cosimo II the heavenly bodies named after his family. The Medicis were not science buffs and were wary of Copernicanism, which they knew might get them in trouble with religious authorities in Rome. But they saw Galileo as a godsend because he had supplied them with things like water-lifting machines, compasses, telescopes, and other instruments of use to the state—and now he was branding parts of the universe for them!

Galileo's self-promotion worked. Three months later, Cosimo II invited Galileo to become "Chief Mathematician of the University of Pisa and Philosopher and Mathematician to the Grand Duke." Galileo had insisted that his title include both "philosopher" and "mathematician." He wanted the first title for its status, to poke at his rivals, and because it accurately reflected what he did: explain the ways of the world. He wanted the second title because it accurately described the skill by which he did so.

Soon after he arrived in Florence, in 1610, Galileo discovered that the planet Venus had phases like those of the Moon, another key piece of evidence confirming the view that not only the Earth but all planets move about the Sun. In 1611, encouraged by the evidence, he traveled to Rome to demonstrate his device and argue for his findings. Some Church officials were initially skeptical, but eventually decided he was right. Galileo acquired admirers in the Church hierarchy, including the influential Cardinal Maffeo Barbarini of Florence, who saw nothing much heretical in his work. Galileo was made a member of the Lincean Academy, established a few years earlier for "the study of the great book of nature."

Back in Florence, Galileo trained students and found positions for them at universities. In 1613, he secured his former Paduan student Castelli a job in Pisa. The head of the university gave Castelli a kind of loyalty oath, asking him not to discuss topics that seemed inconsistent with Church doctrine, most notably the motion of the Earth. To keep the peace Castelli signed on, noting that Galileo, too, followed this advice.

The peace lasted about a month. Then Castelli was caught unaware by an episode that took place after he was invited to the royal palace by Cosimo II's mother Christina.

The occasion began innocuously. Cosimo and Castelli discussed life at the university, and then Cosimo asked if Castelli had a telescope. Castelli said yes, and described observations he had made just the previous night of the Medicean planets. Christina asked how he knew that they were real objects and not illusions produced by the telescope's lenses. Castelli explained why, and after some discussion he had all the guests fawning over Galileo's wonderful discoveries. But as Castelli was leaving the palace, a servant caught up with him and ordered him to return. He was led to a back room where several guests had gathered, not only Christina but Cosimo and a friend who was a philosopher. Evidently the latter had put a bug in Christina's ear to the effect that the motion of the Earth contradicted the Bible. Christina cited some biblical passages and wanted to know what Castelli thought.

Castelli carefully "played the theologian," as he told Galileo afterward, "with such assurance and dignity that it would have done you good to hear me." It was unthinkable, he pointed out, that the Bible could possibly contract the evidence of the senses. Only heretics could claim such a thing! He won everyone over except for the philosopher, who remained quiet, and Christina herself—though Castelli suspected that she was just playing devil's advocate to hear his response. Still, the edgy episode shook Castelli enough to write Galileo a lengthy description.

BACK AT THEM

Galileo was then the jewel of the Tuscan Court. He was not yet fifty, but the news of his world-shattering discoveries—that the Moon had craters and mountains, that Jupiter had satellites and Venus phases—had spread throughout Europe and (thanks to Jesuit missionaries) had reached even China. Besides prestige and profile, he was also supplying the Tuscan Court with militarily useful instruments like telescopes.

His personal life was bumpy. Marina had died, and Galileo was supporting two daughters and a son. A new mistress (Cassandra) had recently given birth to his fourth child (Anna).[9] Galileo was not prosperous enough to support his children at the level expected of someone with his court stature. Social conventions dictated that illegitimate daughters were unmarriageable, so Galileo entrusted their care to the San Matteo convent in nearby Arcetri, where Virginia was renamed Sister Maria Celeste and Livia became Sister Arcangela (nothing further is known about Anna). Virginia and Livia would spend the rest of their lives there. By now, Galileo had his son Vincenzio living with him, and managed to get Grand Duke Cosimo II to have him legitimated. Still, Galileo had enough money to pursue his own scientific inquiries thanks to his instrument business and the court's patronage.

Yet Castelli's news alarmed Galileo. Like other adherents to Copernicus's heliocentric theory, he occasionally faced accusations from biblical literalists that the theory was in conflict with the Bible. He and others made light of the charges, pointing out that the discrepancies were minor and did not contradict the Bible's spirit. So far, Church authorities had agreed, and thought the matter not worth investigating. But news that someone was whispering these allegations directly in the ear of his patron's mother was alarming. The Church had condemned and executed heretics, including the Dominician friar Giordano Bruno just fifteen years earlier. Even if Cosimo II were on Galileo's side, the Duchy was not powerful enough politically to resist the Church. Suddenly Galileo felt that his work, and even his life, was threatened.

The telescope was largely to blame; it had disrupted a fragile truce between Copernicanism and the Church. The claim that the Earth moved around the Sun was incompatible with only a few loose and poetic sentences in the Bible (e.g., Ecclesiastes 1:5: "The Sun rises and the Sun sets, and hurries back to where it rises"). Still, in the sensitive post–Council of Trent climate, even these few sentences could not be overlooked. If they were wrong, it undercut the entire effort to reestablish the Church's authority to interpret the Bible. For a while, showing that one was a good Catholic had been enough for a Copernican to avoid conflict with the authorities. Galileo himself had survived this way, playing the theologian well and ducking questions of the compatibility of his findings with scripture. The telescope now made this impossible. It produced clear evidence that not *everything* in the Bible was *literally* true. Copernicanism was no longer a speculation that convinced you if you did the math; the evidence was obvious. The telescope was the real threat to the Church authorities, not because its discoveries ran counter to moral doctrines, but because it showed that these authorities did not know what they were talking about.

Galileo was determined to find a way to lay the issue to rest. He was an excellent writer—the Italian novelist Italo Calvino judged him the greatest Italian writer of all time—and loved sarcasm and ridicule. Once engaged in a fight, his cleverness, learning, doggedness, and wicked humor made him an intellectual bulldozer. You didn't mess with him unless you were stupid or had an army to back you up. For the Catholic Church authorities, it was both.

In December 1613, days after receiving Castelli's letter, Galileo wrote back with his thoughts on why the new science was compatible with the Bible. He meant for Castelli to share its contents, and circulate copies to friends. This worked for a short time. Then a Dominican friar in Florence attacked Galileo as a heretic in a sermon, citing biblical passages as his authority. Another Dominican friar repeated the attack, raising the stakes by sending a complaint directly to the Inquisition in Rome along with a purloined copy of Galileo's letter to Castelli. It was ominous that Church authorities were now involved. The Roman Inquisition cleared Galileo, but accusations against him continued to circulate.

In the summer of 1615, Galileo decided to meet this challenge head-on, and drafted a letter to make explicit why his astronomical studies and biblical sentences are equally authoritative. Christina seemed the right addressee, because she had hosted the gathering where the issue first surfaced and was the matriarch of his patron's family. But Galileo clearly meant it as an open letter to political and religious authorities. The letter, which he never sent, is neither apology, ruse, nor evasion. It is a forceful and often biting explanation of the authority of science, and why one needs to defer both to it and the Bible. It turned the arguments of his theological opponents back against them.

The letter is long and goes into some fine points of Church doctrine, so I'll paraphrase loosely. A few years ago, Galileo began, he came across some surprising findings about nature, such as the fact that Jupiter has its own planets. Most scientists accepted the findings right away, and a few others did after they saw the evidence. A few people attacked him, but not because they found specific flaws in his work. Instead, these adversaries had accused him of making it up, and of undermining traditional values. A few even refused to look through his telescope.

I fully accept the authority of the Bible, Galileo continued. But those who reject my findings or think that they conflict with scripture do not understand either the findings or the Bible. The Bible is written for ordinary people, not scholars, and it's about morals, not nature's details. It doesn't talk about the solar system at all, for instance, and mentions one and only one planet, Venus, which it rather bizarrely refers to as "Lucifer." As a theologian friend of Galileo liked to say, the point of the Bible is "to teach us how one goes to heaven, not how heaven goes!" The Bible even has some poetic passages—that for instance describe God as being forgetful, or not knowing things he should know, or having certain other all-too-human traits—which if we take literally would make us fools and heretics, as the great theologian St. Augustine pointed out.

The bottom line, Galileo concluded, is that God wrote two books, the Bible and nature itself. To understand the first, we go to religious authorities, while to understand the second, we have to "read" it; that is, we must study it ourselves. These two books cannot contradict each other, for they have the same divine author. If they seem to contradict each other, then we are misreading the one or the other.

Galileo therefore went right back at his opponents. He began by accepting the most fundamental values that they held, or claimed they held. Then he showed that these values did not dictate any specific findings—and that seeking these findings was completely in the spirit of those very values. Just as his *Inferno* lectures had shown that "Dante literalism" was not enough to understand the shape of hell, so his letter demonstrated that "biblical literalism" was not enough to understand the structure of the world. Galileo's letter therefore amounted to a defense, not just of his telescopic findings, nor of the use of math, but of the entire authority of the scientific workshop.

It is hard to imagine how to put together as forceful a defense in today's world because contemporary science deniers do not adhere to a single set of values as Galileo's theological opponents did. Those who deny climate change and evolution, who assert a connection between vaccines and autism or deny one between AIDS and HIV, are not motivated by a single religious system but by an array of motives—greed, fear, bias, convenience, or profits, among other reasons—to which they cling with various amounts of sincerity and cynicism. Still, Galileo's general strategy—to accept right away the fundamental values of one's opponents—is instructive. If, say, the deniers paint themselves as patriotic Americans in opposing global warming, a Galilean way to oppose them would begin by embracing the Constitution as the law of the land and elected officials as its stewards, and accepting such basic American values as jobs and the economy. It would then point out that, though politicians are the ultimate political authorities, one does not consult them for technical advice about practical issues such as glaciology or climate. That's a scientist's job; scientists study the reality of these things, which allows politicians to know what practical actions are possible, how effective they are, and how costly.

When politicians try to usurp scientists in this respect, they simply display their ignorance and humiliate themselves—as when Senator Jim Inhofe (R-OK) tossed a snowball in the halls of Congress in 2015 to "disprove" global warming, revealing his ignorance about the elementary difference between climate and weather. When politicians do so, they and not scientists are the ones ultimately threatening basic American values, such as jobs and the economy. The Constitution may structure the laws of the land, but not the laws of nature. The Founding Fathers showed how to create legislation, not how to legislate creation! There are two Constitutions, as it were, the American Constitution and the Constitution of Nature. Politicians are authorities of the former (or claim to be), scientists authorities of the latter. These two Constitutions—political and natural—cannot conflict. If they appear to do so, somebody is overstepping their authority.

Thus both Bacon and Galileo appealed to the image of the book of nature as a must-read for devout humans. But while Bacon promoted the necessity of the book's message, Galileo pointed to the authority of its language. Bacon used the book metaphor to motivate the King to build a scientific workshop, and Galileo to defend it against attacks from theologians.

THE FALLIBILITY OBJECTION

Galileo's brilliant defense didn't work, at least not in the short term. The next year, 1616, Pope Paul V convened a panel of theologians to study Copernicanism. But scriptural literalists had made too much noise for the panel members to politely avert their eyes. The panel found the idea of a Sun-centered universe heretical because it contradicted the Bible, but due in part to the intervention of Galileo's friend Cardinal Barbarini, the Church did not issue a public and official condemnation. The exact message the Church issued is unclear, but included an injunction not to "hold, teach, or defend" the Copernican view.

For a few years, Galileo had been staying with a friend on the outskirts of Florence. In 1617, he rented a villa in the hills south of the center of Florence, from which he could see both the walls of the city and the convent where his beloved daughters lived. There he continued to work, technically obeying the Church's injunction. But he kept honing his arguments, and in particular the metaphor of the two books.

In 1623, he thought he saw an opportunity to safely reopen the matter when his friend Cardinal Barbarini became Pope Urban VIII. Galileo now wrote *The Assayer*, which critically evaluated or "assayed" the work of a certain traditional astronomer. Galileo dedicated the book to the pope, and prominently displayed the new pope's coat of arms in the frontispiece. Obediently, Galileo does not mention Copernicanism. (The book's main scientific thesis, though, that comets are optical illusions, turned out to be incorrect.) But the book does lay out Galileo's method, and in his most well-known sentence he gives the metaphor definitive form. The world (philosophy), he says, is written in a "great book" that we cannot understand unless we learn its language and symbols. "The book is written in mathematical language, and the symbols are triangles, circles and other geometrical figures, without whose help it is impossible to comprehend a single word of it; without which one wanders in vain through a dark labyrinth." Pope Urban, delighted with the book's wit, insight, and sarcasm, had it read to him at mealtime.

Galileo had stood the old metaphor on its head. In medieval times, and even for Bacon, the two "books" had been meant to be read together, each comprehensible to the layperson and serving as a guide to understanding the other. Galileo was now implying that the book of nature had a deep, self-contained meaning accessible only to the mathematically literate who could read its special language. Galileo's defense was ingenious and powerful. God, he said, had used mathematics in his blueprint for creation—in writing the world's technical manual, so to speak—while Scripture was a kind of users' manual, written in loose, popular talk. Scientists reading the mathematical "text" of the book of nature are therefore speaking God's own language. Their authority was therefore as solidly and divinely grounded as that of the ordained priesthood.

When Galileo next visited Rome, the pope led him to believe that he could write a book on Copernicanism if it fairly laid out the arguments pro and con. He also insisted that Galileo admit something else—that because humans could never know the mind of God and all of His activities, scientific knowledge can never be fully complete. Galileo agreed with this, what we might call "Urban's Principle." Galileo set to work, writing the preface with a censor for safety in which he said he would treat the Copernican system as a theory only, and that religiously inclined readers were free to choose the Ptolemaic system. Galileo also wrote the book as essentially a play—an extended dialogue between three fictional people: Salviati, a stand-in for Galileo and proponent of Copernicus; Simplicio, an ideologue (modeled in part on Cremonini) who advocates the Ptolemaic system; and Sagredo, an intelligent laymen.[10] Galileo also has Salviati note that the discussants are merely trying to enlighten each other, and that the issue is for other authorities to settle.

Simplicio makes an interesting observation on the second day of the dialogue, when he complains that they have been talking about abstract things, not real and concrete bodies. But he is promptly knocked down by Salviati. That doesn't matter, Salviati replies. To understand real bodies and events you have to think abstractly, like in accounting. When businessmen track sugar, silk, and wool they leave out details regarding containers, straps, and other incidental packing items. Similarly, to understand the motions of objects scientists have to "deduct the impediments of matter," like friction, stickiness, and other things.[11] Science, that is, requires a certain technical skill. But Galileo made it sound easy. On the fourth and final day of the dialogue, Simplicio voices Urban's Principle—that because no scientist can imagine the whole of God's "power and knowledge," any specific conclusion by humans about His Creation is uncertain. It is sometimes called the "divine omnipotence" or "fallibility" objection to scientific certainty.[12]

Despite Galileo's precautions, the *Dialogue Concerning the Two Chief World Systems*, as it was called, was too provocative and disobeyed the spirit of the 1616 injunction and what had been expected of him. It did not treat Copernicanism as a mathematical speculation but as a real possibility for how the world worked. It also portrayed the authority of the senses as superior to book learning, yet showed that the testimony of the senses (as in the Sun "rising" and "setting") could also be untrustworthy.[13] Furthermore, it was not smart of Galileo to place Pope Urban's deep thought about the ultimate fallibility of science in the mouth of Simplicio, the fool. A final contributing factor was that Urban's influence within the Church—his ability to defend Galileo if trouble occurred—was weakened by mounting debt created by the Thirty Years' War. All these pressures made Urban, Galileo's former friend, feel he had no choice but to turn the matter over to the Inquisition. Church authorities ordered Galileo to Rome, threatening him with arrest if he refused.

Galileo's trial took place in Rome in 1633, a year after the *Dialogue* was published. He defended himself vigorously, but lost anyway. Though he apologized for violating the terms of the 1616 judgment and publicly recanted heliocentrism, he was sentenced to house arrest. That he renounced the recantation under his breath, muttering "eppure si muove" or "it moves anyhow," is a legend.

House arrest meant moving back to his villa outside Florence, in an area called Arcetri. He had moved there in 1631 to be closer to his daughters, who were in the nearby Convent of San Matteo. Over the remaining nine years of his life he stayed in Arcetri, in the company of his daughters and visited by celebrities including the poet John Milton. Galileo wrote one more book, the *Discourses and Mathematical Demonstrations Relating to Two New Sciences*, known as the *Two New Sciences*; though it was about the laws of motion, he avoided discussing the motion of the Earth.

GALILEO AND SCIENCE DENIAL

Galileo's greatest contribution to humanity did not involve telescopes, pendulums, or mathematics. His most revolutionary act was to help usher in a way of looking at the world in which we *can* look at the heavens with telescopes, think of its chandeliers and swings as pendulums, and read its book in the language of mathematics. This was the "crazy universe" hinted at by Abraham Pendulum, that Galileo explored in Pisa, Padua, and Florence, and elaborated in the *Dialogue* and the *Two New Sciences*. This universe would soon be fully charted by Newton, who was born the year Galileo died.[14]

Galileo's defense of the authority of those who understood this world was brilliant, but it inadvertently exposed two potential vulnerabilities about science.

The first vulnerability was the gap between the scientific world and the world of ordinary human affairs. The emergence of Galileo's gap, as I call it, is a moment of revolutionary change in human history, more significant than any war or political development. In their professional work, the experts who ran Bacon's Bensalem would have to think differently from the citizens on the rest of the island—they would speak another language. (The next chapter, on Descartes, will explore this more thoroughly.) The rupture between these two ways of thinking was the dawn of modernity.

That gap has since widened and grown larger in size and consequence. After Galileo, that ordinary world, which depended more and more on the scientific world, became harder to inhabit. It needed informed inhabitants who knew how to cross that gap, to travel back and forth from that abstract universe to our own. The danger arose that some people would be unable to make that transition, and become stuck on one side or the other of the gap, staring incomprehensibly at those on the other side. The danger arose, in short, of an apartheid between mathematicians and scientists on the one hand, and everyone else on the other. The danger arose because members of the general public were getting the impression that the workshop is inhabited by a scientific priesthood and that ordinary people are locked out, not given its passwords. That impression is one factor that enables modern-day Cremoninis to protect their views by saying, "I am not a scientist," and refusing to look at the technical evidence that challenges their values.

The second vulnerability is that science can never be complete. Pope Urban, after all, was right. Part of the strength of science is its perpetual openness to revision, but that feature seems to give legitimacy to claims—such as those made by former Environmental Protection Agency (EPA) head Scott Pruitt at his confirmation hearings—that "the jury is still out" on important issues such as climate change.

Still, Galileo's story also shows how to respond to such claims. It is not enough to respond by saying things like "science works" or "science is epistemologically justified" (meaning, it's logical). It is necessary to be more aggressive. Science deniers have to be made uncomfortable by shoving back in their faces the brute fact that their claims contradict the very values they say they hold dear. One has to aggressively appeal to the sources of authority accepted by the deniers themselves.

THREE

RENÉ DESCARTES: WORKSHOP THINKING

I ONCE ATTENDED a contentious meeting about the public health impact of exposure to low levels of radiation. One speaker was a scientist who carefully reviewed the research statistics: how the data had been gathered, how they had been reviewed and confirmed, and how they fit into what else was known about health effects. An antinuclear activist in the audience stood up, shouting, "You love the numbers more than you love people!" A vast majority of the audience fervently applauded. They clearly thought that the scientists present were callous and wanted them silenced. I felt a shiver of fear.

The scientist was shocked into silence, but then he told a little story. A few years ago, he said, his son asked him if it was safe to install air bags in the car to protect his child—the scientist's grandson. Newspapers had run horrific tales about incidents where air bags had smothered children, accompanied by grisly photos. Concerned, the scientist had looked up the scientific studies on air bag safety. The statistics, he found, revealed that it was much safer to install the air bags, which he advised his son to do. The scientist concluded, "I love the numbers because I love my grandson."

The quiet humility of the scientist calmed the crowd somewhat, for a while at least. I never forgot this, because it illustrated the enormous difference between thinking inside and outside the workshop, as well as the suspicion with which workshop thinking can be perceived by the general public. It also shows that genuine public welfare needs both workshop thinking—investigating the numbers—and regular thinking.

The French philosopher René Descartes (1596–1650) provided the first detailed description of workshop thinking and expertise, and its difference from ordinary thinking. Descartes realized how great the difference was, and that going back and forth could be a struggle. His description would be the first description of workshop expertise.

MR. EVASION

Descartes, who had an exceptionally independent and impervious personality, spent a lot of time thinking about thinking. To say that he was "in his head" only scratches the surface.

He came into this world virtually an orphan. Descartes's father, a lawyer, government official, and nobleman, was away at a Parliament meeting when his son was born on March 31, 1596; his mother died a year later giving birth to another son. At age four, Descartes was left with his nurse and grandmother in provincial France when his father remarried and moved away.[1]

Descartes's birthplace, a small house in the town of La Haye, south of France's Loire Valley, is now a museum. It has only a few artifacts: a plaster cast of his skull, a few musket balls from local wars, and a life-sized mannequin of the philosopher holding a pen. There is a permanent exhibit of posters that explain the religious, political, social, and health turmoil that buffeted France in his lifetime, costing thousands of lives and nearly bankrupting the government. But the exhibit also shows how much the 1500s in Europe was a time of exciting discoveries and achieve-

Descartes's birthplace.

ments. The Europeans had only recently learned about the Americas, and had circumnavigated the globe about seventy years before Descartes was born. The young introvert was entering a tumultuous age—people were still adjusting to the implications of these revolutionary events and a new age of discovery.

La Haye was a farming community that has since been renamed Descartes. Living with his maternal grandmother, Descartes was curious and self-motivated. He enjoyed wandering the countryside by himself and was fascinated by farm implements. When he came across a particularly ingenious one, he wondered how he himself might have invented it. For his entire life, he derived pride and pleasure from doing things on his own.

At age eleven, in 1607, Descartes was sent to an eminent Jesuit boarding school nearby called La Flèche, founded by Henry IV a few years earlier. There his education included Greek and Latin classics as well as contemporary scholarship and science; there he learned of Galileo's telescopic discoveries. But he was most interested, he wrote later, in things "useful to life." He got perks denied most other students, thanks to his well-off family and a relative who worked at the school. He was allowed to keep his reclusive habits. While classmates were summoned by a 5 a.m. bell, Descartes was allowed to sleep late and was even given his own room. From childhood on, he spent so much time shunning human company that one person nicknamed him Monsieur d'Escartes: "Mr. Evasion."

In 1610, Henry IV was assassinated by a Catholic fanatic, an act that indicated ongoing religious tensions. The fourteen-year-old René, along with other students, attended the local memorial service.

René's father wanted him to become a lawyer, and in 1615, after La Flèche, Descartes studied law at the University of Poitiers, another nearby institution. But law disappointed him for the same reason that Aristotle had disappointed Bacon; it seemed to inspire people to bicker about words rather than invent useful things. When Descartes turned twenty-two in March 1618, he quit Poitiers, stopped studying the law, and left tumultuous France. Wanting to become a military officer and see what he called the "great book of the world," he joined the army of the Dutch Republic. The Dutch Republic, though Protestant, was a French ally in the Thirty Years' War against Spain. Many artists and philosophers found refuge in wealthy and tolerant Holland.

INTELLECTUAL AWAKENING

Descartes's intellectual awakening took place that fall when he was stationed in the town of Breda. A temporary truce with Spain gave him free time. Descartes met a visiting school official and amateur scientist named Isaac Beeckman (1588–1637), and the two argued over a fine point of mathematics. Beeckman set the cocky young French soldier straight, but the two bonded. It was not only because, Beeckman wrote later, they were the only two people around who spoke Latin. They also shared a conviction that mathematics is deeply entwined with the world.

Descartes's first biographer recounts that the initial meeting between the two was in front of a poster that advertised a public mathematical challenge—a not unusual practice at the time. The poster was in Flemish, which Descartes did not know well, and he asked Beeckman to translate the problem into Latin. Amused by a soldier's interest in math, Beeckman was surprised when the youth claimed he would shortly solve the problem, and astounded when Descartes later showed up at his house with a solution. This is the sort of suspicious detail (a biographoid, we may call it) that historians hate. It is surely embellished, has no supporting evidence, and suspiciously resembles stories told about other people, but it contains just enough grains of truth to make retelling it irresistible. Similar biographoids include the story of Bacon dying after stuffing chickens full of snow, and Galileo muttering "Eppur si muove" after his condemnation. The grains of truth are that Descartes was young and arrogant while Beeckman was older and clear-headed, and that both were interested in science and math and finding things that were useful to life.

Descartes helped Beeckman develop mathematical skills, while Beeckman showed Descartes how to apply mathematics to physics. The two had an intense friendship until 1629, when they quarreled; Descartes did not like that Beeckman viewed him as a disciple, and Beeckman grew tired of the younger man's claim of how original all his insights were.

In 1619, Descartes had a dream that he thought revealed to him "a marvelous science" in which all knowledge was linked together systematically, as if bound by chains. Descartes spent the next nine years trying to construct this science, crossing Europe to collect information. At the time, European scholars were linked in a long-distance community known as the "Republic of Letters." The key institutions in this republic were universities, and the main mode of communication was the postal service; scholars wrote letters to each other in Latin that were circulated to local contacts. Descartes's travels resulted in his meeting personally many of the most important and widely connected scholars of the day. These included the philosopher and mathematician Marin Mersenne (1588–1648), the networker-in-chief of an influential group of scholars in Paris.

But Descartes spent much of his time alone, and often changed houses within each city to protect his privacy. In 1629, he settled in northern Holland, where he sought to build a machine to grind lenses for telescopes, and to develop his comprehensive science.

He never built the lens-cutting machine, but he made great progress on putting together his science, in which all knowledge was linked. What was breathtakingly ambitious was his assumption that science involved conceiving *all* nature operating mechanically according to the *same* laws, which would help to usher in the modern practice of proposing and testing models. It turned into one of the most ambitious and influential projects ever tackled, rivaling Bacon's *Great Instauration* and Galileo's *Two World Systems*. Descartes simply called his *The World*.

CREATING *THE WORLD*

The project began with Descartes's desire to explain parhelia, the bright spots of light dotting the ring around a solar halo like brilliant beads. But he found he couldn't complete this project without getting into other issues. Soon, he wrote Mersenne, he "decided to explain all the phenomena of nature, that is to say, the whole of physics."[2] Bacon and Galileo had studied the foundation of the sciences, but Descartes found their approach rather scattershot. Aiming at something more systematic, he thought that their work had to be done over from the ground up. Descartes felt that his methodical procedure gave him the key.

Unfolding these thoughts required more research. For several months in Amsterdam, Descartes visited a butcher almost daily, watching him carve up animals and ordering organs to personally dissect at home. Townspeople mocked him for literally getting his hands dirty for science. But he learned more about animal anatomy than was known up to that point, and was able to explain how most animal parts worked. He would write that the parts even of the human heart—valves, vessels—are as integrated as those of a mathematical demonstration, and challenged anyone who doubted him to witness directly the dissection of the heart of a large animal. "No other great philosopher, except perhaps Aristotle," states the *Dictionary of Scientific Biography*, "can have spent so much time in experimental observation."[3]

Descartes and Aristotle, however, viewed things completely differently. Aristotle pictured the world as composed of different places (Earth and heavens) populated by dissimilar substances (on Earth, natural things and human creations) obeying their own laws. But *The World* and its companion pieces envisioned a *single* universe of mechanisms, from plants and human bodies to the Sun and planets, operating according to the same principles. The human organs are all machines. The heart is a pump, and the nerves hollow tubes through which a water-like substance flows to operate the body "like the mechanical statues found in the grottoes and fountains in the gardens of our Kings." The nerves relay stimuli to a command post in the brain (the pineal gland, Descartes speculated, because it was the only part of the brain for which he could not find a double) where the conscious part of human beings receives and issues messages. The rest of the natural world—from sticks and stones to the Moon and planets—would also behave mechanically. Descartes envisioned space as filled to the brim with tiny particles called corpuscles; when something moves it pushes aside the corpuscles in front of it while others rush in to fill the space it had just left. Swirls of corpuscles called vortices provide the powerful forces needed to keep the planets moving. "I have described," he writes toward the end of his *Principles of Philosophy*, "the whole visible world as if it were only a machine in which there was nothing to consider but the shapes and movements [of its parts]." The scientist's job was to work out the engineering.

Descartes was not alone in seeing mechanical movements in nature; in 1605, Kepler had written that "the celestial machine is to be likened not to a divine organism but rather to a clockwork." Descartes, it turns out, got many of the world's laws and mechanisms wrong, including the idea that space consists of material corpuscles, but he helped to usher in a mechanistic way of thinking.

Part of Descartes's vast cosmic machine was the Earth's motion around the Sun. Descartes knew of the Church's 1616 warning to Galileo about heliocentrism, but he did not take it that seriously. He had heard that people in Rome taught heliocentrism anyway, and thought the cardinals who had objected would soon change their minds.

Still, Descartes remained wary of how the Church would judge his work, and framed *The World* (and the *Treatise on Man*, intended to be published together with it) as a thought experiment: imagine that God created statues operating in this mechanical way, each linked to a soul via a gland in the brain: wouldn't these be just like us? What if God created a "new world" in "imaginary space" in the form of a giant machine with corpuscles and vortices: wouldn't it be indistinguishable from our world? Descartes invited readers to adopt a remote and detached perspective on the world. Think, he said, of the way sailors on a boat far from land view the ocean: it seems to extend infinitely in all directions and one can take it all in at once. The image expresses an early understanding of "objectivity" as an aloof and almost God-like perspective.

GALILEO'S CONDEMNATION

In the summer of 1633, Descartes was living in Utrecht, completing his ambitious work and hoping to send his friend Mersenne a copy by New Year's. But that November, Mersenne wrote him that Galileo had been condemned and barely escaped with his life. He told Descartes—incorrectly, it turned out—that all copies of Galileo's *Two World Systems* had been burned in Rome. Descartes was terrified. "I was so astonished at this that I almost decided to burn all my papers," he wrote Mersenne that month.[4] He didn't know the details, but realized that the condemnation had to be due to Galileo's heliocentrism. Descartes was living in the Netherlands, a Protestant country out of the political reach of the Catholic Church authorities in Rome. Like Galileo, however, he accepted the Church's authority in spiritual matters, even as he sensed an authority of another kind. He told Mersenne that while he had "very certain and very evident demonstrations, nevertheless I would not for anything in the world maintain them against the authority of the church."

But the chain-like unity of his science that had so inspired his dreams now turned around to haunt Descartes. Its elements were as internally connected as those of math. Just as you can't take the Pythagorean theorem out of geometry and still have geometry, so you couldn't take the heliocentrism out of *The World* and still have the world. "If the view [heliocentrism] is false," he continued in his letter to Mersenne, "so too are the entire foundations of my philosophy, for it can be demonstrated from them quite clearly." He hoped the condemnation would be reversed. Still, he preferred to suppress his book rather than publish something "in which a single word could be found that the Church would have disapproved of." This was neither an atheist's posturing nor a devout believer's humble acquiescence. It was a political decision, stemming from fear that the exercise of raw power would endanger the work he loved. His greatest desire, he told Mersenne, was "to live in peace and continue the life I have embarked on, taking as my device the motto: *he lives well who hides well*."[5]

Descartes knew how to hide well, and had carefully masked his personal life even from close friends. In summer 1634, he began a year-long affair with Hélène, the servant of a bookseller friend, who gave birth to a daughter, Francine, the following year. This is Descartes's only love affair that historians know of—and they know little. He appears to have been a good father, and to have been brokenhearted when Francine died five years later. Years later, when an enemy accused him of having illegitimate sons, Descartes righteously denounced the technically groundless accusation, though he added, "I am a man and did not take a vow of chastity, and never claimed to pass as better behaved than other men."[6]

Descartes was in a Protestant country, so Galileo's condemnation did not make Descartes fear for his life. Still, it sent him into spiritual crisis. In areas under Catholic political control you could be burnt at the stake for choosing wrongly. Yet, like Galileo, Descartes knew himself to be both a believing Catholic and a Galilean. He accepted the Church's moral authority, but needed the mental space to be a scientist as well.[7]

He found a way. Descartes decided to publish three essays that exemplified his new science but were not closely dependent on heliocentrism. One was a mathematical essay containing key inventions, such as the familiar "Cartesian" graphing system you learn in high school that specifies points by their distances from each of two perpendicular axes.[8] Another essay, on optics, mentions plans for the lens-cutting machine and includes the first published version of the law of refraction.[9] The third essay, on meteorology, updated Aristotle's work on the subject, discussing mechanical models of things like snowflakes and rainbows.

Descartes then wrote a preface to explain and defend what he had done. Known as the *Discourse*, it is his description and defense of workshop thinking, and one of the finest pieces of philosophical writing ever. He wrote in French, the language of his countrymen, rather than in Latin, because he wanted to rely on those who used their everyday understanding rather than the convoluted reasoning one learned in school. He published the work in the Netherlands anonymously, so people could decide for themselves whether or not the material was heretical. One of its main accomplishments was to describe a way of thinking that could be legitimately conducted apart from any questions of what the right religion was. Let's call it "sequestered" thinking, the type of thinking that you do momentarily apart from the rest of your life, a mental retreat. By describing such thinking—what it consisted of, what made it different from other thinking, and what its value was—Descartes hoped the preface would be persuasive enough to eventually enable him, as one friend put it, "to bring his *World* into the world."[10]

EXPERT THINKING

Descartes—or rather, the anonymous author—begins with a little joke. Everyone must have the same amount of common sense, he says, because nobody complains of not having enough. But it's a deep joke. It suggests mental democracy: when people disagree it's usually not because they are mentally defective, but because they have misused their minds or become distracted. It says, in effect, "I've found how to stick to the straight and narrow, to keep from mentally going 'on holiday'; that improves my grip on the world, in certain areas at least, and I'll show you how I found it."

In the first part of the *Discourse*, Descartes describes his disappointment with his education. Though he attended one of the best schools in Europe, all he learned was how little he knew. After he quit, he spent at least one entire day alone planning an educational strategy. He decided he would have to go it all alone, finding a solid foundation and using math as a model. This meant proceeding in four steps: (1) using only pieces of knowledge he knew to be solid, (2) beginning with the smallest pieces, (3) assembling them one by one, and (4) checking his work early and often. Meanwhile, he would obey the laws and customs of whatever country he was in to avoid distractions and interference. Following this plan, he could live amid the hubbub of social life while still having the leisure and focus to examine his ideas and opinions. He decided to cast out any that seemed challengeable or poorly established—in fact, he would go so far as to pretend that such opinions were false. He didn't *really* think they were false. He was just trying to see whether there were any ideas and opinions he absolutely could not throw out.

He found one.

Try saying to yourself—and meaning it—"I am not now thinking." Descartes tried, and couldn't do it. If you think, you are. That's not a logical truth. Nor is it a fact that you just discovered—you already know you exist! It's an experiential truth, Descartes realized, that you find when you clearly and distinctly grasp your own actual thinking at a specific time and place. He had found a truth that no politician, no theologian, no authority—not even a God—could convince him is false. This doesn't mean he doubted God, or questioned His authority. It only means that one's other beliefs are irrelevant to this particular truth. A Christian, Muslim, or madman can get it. Religion and morals are irrelevant to certain kinds of truths, and this is one.

"I think, therefore I am" was the first truth Descartes found in his mental retreat, but he soon found others. For one thing, the experience gave him a criterion for truth: something you clearly and distinctly grasp. He also thought he could then prove the existence of a benevolent God. This sounded backwards; most thinkers started with God and worked from there, while Descartes started with his individual self and then proved God's existence. But that's how he wanted to proceed following his systematic, logical plan. What mattered to Descartes most was to take small steps based on what was right in front of him. That way, he didn't have to worry about theology, politics, and everything else that was tearing apart France and the rest of Europe. If you start from clear and distinct ideas and make sure the results knit together as they do in mathematics, you can do solid science without being heretical or unpatriotic.

It would be nice if science were completely like geometry, something one could work out entirely in sequestration, deriving the principles of each natural phenomenon from higher principles the way one proves mathematical theorems. But God is so infinitely greater than anything we can think or imagine, Descartes wrote, that we cannot possibly understand divine purposes, and thus be completely certain of any particular scientific law. This is Descartes's more explicit and philosophically worked-out version of Pope Urban's thought about the ultimate uncertainty of any scientific conclusion, which I call the "jury is still out" principle. Therefore, you always have to test and experiment in establishing scientific laws in order to find what actually connects to what.

At the end of the *Discourse*, Descartes wrote of how he was about to publish a summary of his findings when he learned that the Catholic Church had condemned Galileo's heliocentrism, which was part of his system. He had objected as a scientist but obeyed as a man, and did not publish. He had changed his mind, and now felt morally obliged to go public concerning his still-unpublished system. He still saw nothing in it that should be objectionable to true religious authorities. If he remained quiet, he would be violating a moral commandment to enhance the general good of all mankind, for his approach describes how humans can acquire the tools to become "the lords and masters" of nature.

Descartes thus laid out in clear and nontechnical language how science works: math structures the principles, experiments provide the details, and all can be understood and utilized by anyone regardless of religion. His defense of the workshop differed in style and substance from those of Bacon and Galileo. Descartes addressed the *Discourse* to lay Europeans, and it was not preachy or rebuking but simply a story. Follow it carefully, he declared, and you will understand why I reason the way I do, and be able to do it yourselves.

SEQUESTRATION

Descartes' *Discourse* describes two key spheres of human action. One is the everyday world of human engagements and interests. We inherit this world, are usually absorbed in it, and have little or no control over it. This is the lifeworld, in the terminology of later philosophers: ever changing and ephemeral, it cannot be mathematically represented or scientifically captured.

In the midst of this everyday world, though, it is possible to build workshops where we can think and experiment in special ways. Inside them, we can be in nearly complete control, almost like the way we are when doing math. This does not mean we are rejecting the lifeworld, only that we are working in a sphere where its involvements will not get in the way. Ordinary thinking can be curious, distracted, or meditative; it does not always lead to results or resolve anything. Workshop thinking, on the other hand, is focused, has a definite aim, and seeks to come to an end. It requires rigorous training to think this way systematically. But if your aim is to come up with something useful, you have to.

One of my students once called the mental training Descartes describes "detoxing." It's a good metaphor in that it captures the difficulty of retraining your thinking, but it's a bad one in that it likens switching to scientific thinking to ridding yourself of a poison. Descartes did not see it that way. He knew well the pleasures and purposes of ordinary lifeworld thinking, and that workshop thinking was never far from it.

The *Discourse* defends science by carefully describing why the two spheres of human action, the lifeworld and the workshop, do not conflict. Here lies both the power and danger of Descartes's thinking. He made science and religion compatible by assigning them to different worlds, one nested in the other. The workshop is a sequestered part of the world, not a detached, atheistic, or abstract hole in it. You can compare the workshop to a sequestered jury, or a group of people who are placed for a short period in isolation. The danger is that, without intending to, those in the workshop will begin to see it as the "real" world. But that would be like jurors thinking that all of life's decisions should be decided when sequestered.

Descartes was proud of the *Discourse* and sent it to prominent scholars. He valued their reactions, within limits. Mersenne had trouble with the informal language and fired back questions on metaphysical issues; Descartes then reminded him it was a preface to the rest of his science rather than a treatise. Others, who tried his patience with obtuse objections, he compared to "flies that buzz around a man's face."[11] Exasperated by their failure to understand, Descartes wrote one friend of his "gray hair that is coming in a rush."[12]

Descartes decided that he had invited many of these misunderstandings because he had written *Discourse* informally for the public. Specialists needed their own rendition. In 1641, therefore, he finished a technical version, the *Meditations*, in Latin, the scholars' mother tongue. It describes more carefully how to pass from everyday to scientific thinking, clarifying metaphysical distinctions that he thought necessary.

MEDITATING

Apart from being in Latin, the *Meditations* has many similarities with the *Discourse*. It is written in the first person, comes in six parts, and entreats people to read it carefully or not at all. It traces a path similar to that of the *Discourse*, with the final meditation describing the human body as "a sort of machine" that works as mechanically as a "clock composed of wheels and counter-weights," except one intimately linked to a soul. Even more explicitly than in the *Discourse*, it makes clear how each step builds on what is learned in previous steps.[13] This time, Descartes tried to head off the kinds of criticisms that had greeted the *Discourse*. He gave a prepublication copy to Mersenne and told him to circulate it to key scholars and ask for their objections in writing—and also to say that Descartes planned to publish their objections together with his replies. To his annoyance, most scholars read the *Meditations* as a set of theories rather than as a series of guidelines for scientific thinking—about the practical experience of what it is like to suppress everything but what is needed to think scientifically. Some scholars, for instance, asked how Descartes could know that other minds exist. But he already knows that other minds exist; he's writing for them, and he knows they can follow him because they all think the same way (remember the joke at the beginning of the *Discourse*).

Descartes was particularly annoyed by questions on his view of the relation between the body and the mind—a misunderstanding that has plagued his public and philosophical reputation ever since. The French philosopher Pierre Gassendi asked how body and mind could possibly be united when Descartes described them as different substances. Descartes replied that he was not worried about how they are linked because he *experienced* them as connected, even if he can *conceive* them as being separate. Descartes was so irritated by Gassendi's remarks that he prefaced his replies with a statement that might be paraphrased as follows: you were surely making such idiotic remarks to remind me that nonphilosophical readers are going to say off-the-wall things, and I thank you for giving me the opportunity to head that off.

Gassendi's objection illustrates a danger soon to become a major impediment to understanding Descartes's work—namely, that the workshop tends to overshadow the lifeworld in which it is nested. Descartes thought his experience of the union of the body and the mind was sufficient evidence that they were united—but such evidence was not recognized in his description of workshop methods. Because his readers were awed by these methods—as we are, when we are absorbed by neuroscientific explanations of everything from love and musical ability to political affiliations and criminal behavior—they did not believe his repeated assertions that the body and the mind are "really and substantially" united. Why not? His readers were reading Descartes from the outside, so to speak, paying attention only to the logic of the concepts he developed. They were not reading from the inside, paying attention to the experiences that led him to develop the concepts, and the ends to which he wanted to put them.

At the end of Descartes's life, a plucky Dutch theology student named Frans Burman cornered the reclusive philosopher in Egmond, a town in north Holland where Descartes was residing. Burman arrived with copies of Descartes's writings, and the two conversed in Latin about seventy passages that Burman had earmarked, with the youth writing down both his questions and Descartes's replies. Among them was how the soul could possibly "intermingle" with the body if they are indeed two different substances. Descartes responded just the way he had to Gassendi, Regius, and later to Princess Elisabeth of Bohemia: "This is hard to explain, but here our experience is sufficient, because it declares the fact so loudly that we simply cannot deny it."[14]

Later in his meal with Burman, Descartes grew annoyed with what he regarded as the student's nitpicking of metaphysical points. Just grasp the issues in a general way and move on! If you get too tangled in metaphysical issues you will get distracted from the important thing, Descartes told the student, which is the proper procedure for seeking things that are useful and beneficial to life. The important part of his writings, Descartes felt, was their characterization of how to think scientifically, not metaphysical issues that spin off from that characterization.

At the time, Descartes was being recruited by Queen Christina of Sweden. Christina was well-educated, a freethinker, and rejected traditional roles for women; she scandalized her court by refusing to marry. Christina wanted to attract scientists to Sweden, declaring that her aim was to make Stockholm the "Athens of the north." Descartes resisted her entreaties for a while. But shortly after his meeting with Burman, he agreed to come tutor her and establish a scientific academy in her country. Descartes arrived in October 1649, on a ship that Christina had arranged to pick him up along with 2,000 of his books. He soon regretted the move. Not only was it that he disliked having to rise early in the depressing cold and darkness to tutor her at 5 a.m.—the only time she was available—but also it turned out that Christina did not especially like his mechanical philosophy.

In February 1650, Descartes caught pneumonia and died. He was initially buried in Sweden, but his bones were exhumed and returned to France in 1666.[15] By this time, Descartes was so worshipped as a thinker that these bones were regarded with the same reverence as religious relics, with his skull and one finger detached from the rest of the skeleton.[16]

THE WORKSHOP AND ITS DANGERS

Included in the Wallace Collection in London, a small art collection open to the public, is a sculpture entitled "René Descartes Piercing the Clouds of Ignorance." The sculpture, done by Robert Guillaume Dardel about 1781, shows the French philosopher struggling to free himself from thick, enveloping clouds. His path out is illuminated by rays of sunlight emerging from a hole in the midst of the clouds. Dardel's sculpture casts Descartes, who played a foundational role in both describing and using the scientific method, as a triumphant liberator.

Though Descartes is a pop-culture celebrity for his remark "I think, therefore I am," scientists have scorned him for missteps. In his 2015 book *To Explain the World*, the Nobel Prize–winning physicist Steven

Weinberg listed some of them: Descartes thought that the Earth is prolate (that its diameter across the equator is smaller than from pole to pole), that vacuums can't exist, that space is full of material corpuscles, and that the pineal gland is the seat of the soul.[17] "For someone who claimed to have found the true method for seeking reliable knowledge," Weinberg wrote, "it is remarkable how wrong Descartes was about so many aspects of nature . . . his repeated failure to get things right must cast a shadow on his philosophical judgment."

"René Descartes Piercing the Clouds of Ignorance," Statuette by Robert Guillaume Dardel (1749–1821).

How, then, can Descartes deserve to be portrayed as an agent of enlightenment?

The answer lies in the clouds—those from which Descartes emerges in Dardel's sculpture. The clouds symbolize the continuing influence, in Descartes's time, of both Aristotle and the Church. During an age of continuous and often savage religious warfare, Descartes showed that practicing science was not heretical. He defended its authority, not by arguing that it promoted religious values as Galileo had, but by showing that its practice would enhance the lifeworld. Weinberg is correct to point out flaws in Descartes's science, but his criticism springs from a mechanistic way of thinking that Descartes promoted and helped legitimate.

Descartes has been read in two very different ways. The way that Descartes is all-too-frequently taught in survey courses is to look only at his concepts. This version of Descartes thinks that the mind and the body cannot interact, has trouble knowing that other minds exist, and supposes that philosophy aims to develop a disinterested, detached, disengaged perspective—a "view from nowhere," in later philosophical language—on the world. This caricatured Descartes, in short, has moved from the lifeworld into the workshop for good. The more philosophically sophisticated way to read Descartes is to look at what motivated his concepts and distinctions, and at the ends to which he was putting them. This Descartes is motivated, as the philosopher Robert Scharff notes, "by the desire to *retrain his own mind*—that is, to discard the orientation of an ordinary believer for that of a disciplined knower."[18] The *Meditations*, Scharff says, is the first description of expert training. Descartes was aware that such training can never be perfect.[19] Because humans are always "intermediate between God and nothingness," he wrote, we have to be constantly vigilant and rely both on innate and empirical ideas. This is the "jury is always still out" principle.

Still, Descartes described the vision of a retrained thinker inside a workshop so carefully and vividly as to inspire vain hopes that the workshop perspective is how people should think about everything, even when they ponder moral values or the existence of God. Today, the image has grown all the more inspiring and dazzling with contemporary instruments and technology: nanotechnology enables us to build materials, genetic engineering permits us to rebuild ourselves, and neuroscience allows us to model ourselves from the inside out.

Caring for the numbers is not for scientists only. It's vital for ensuring health and welfare, and for getting a grip on the modern world. The residents of Flint, Michigan, whose drinking water was contaminated by lead in 2014 but whose city, state, and Federal officials steadfastly ignored indications of severe health problems, need to know the numbers—both of the actual and the safe levels of lead in their drinking water. So do West Virginia residents who live near the Elk River, which was severely polluted by a chemical spill also that year. Those who care about the health of Chesapeake Bay and its waterways need to know the numbers indicating the amounts of nitrogen and phosphorus runoff from densely populated areas. The inhabitants of Fukushima Prefecture in Japan, which was exposed to radioactive material after an accident at several reactors at the Fukushima Daiichi Nuclear Power Plant, need to know the numbers—both of the actual and safe exposure levels in their environment. Anyone truly interested in things like melting polar ice, rising sea levels, ocean acidification, vanishing coral reefs, vanishing of organisms that depend on those reefs, migrations and extinctions of species, and glacial melting needs to know the ppm of carbon dioxide in the atmosphere—the actual amount, the change over time of that amount, and the impact on the atmosphere of changes in that amount.

In his description of the mechanical way of understanding the world, Descartes outlined the value of caring for the numbers. The force and vibrancy of his description could make it seem as though he was arguing that that's the only way to think, and therefore that those who think mechanically and care for the numbers have lost their ability to relate to the world. More careful attention to his work shows his appreciation of the importance of thinking both inside and outside the workshop. Descartes's work is therefore important not only for the way it outlines the difference between workshop and ordinary thinking, but also for exposing potential problems in their relationship.

II

At the beginning of the seventeenth century in Europe, efforts to understand nature changed dramatically in substance and tone. In what is generally called the scientific revolution, these efforts resulted in the most radical change in the way humans thought and lived since agricultural methods were developed in the Neolithic era some 10,000 years ago. Bacon, Galileo, Descartes, and others found that passive observing and recording was not enough, and that understanding nature required actively interrogating it and intervening in natural processes.[1]

Each of the trio believed in God and in a God-created cosmos. Each would have agreed with the Dutch biologist Jan Swammerdam (1637–1680) that one can see "the Omnipotent Finger of God in the anatomy of a louse." But each also thought that, in the midst of this cosmos and without disrupting it, humans could build specially equipped workshops for making discoveries and inventions that would improve human life. Furthermore, each regarded creating such workshops as carrying out God's will, not undermining the prevailing Christian belief.

Collectively, their works showed that scientific workshops had a new kind of authority. The people in them were guides to the contours of nature. These guides could be consulted in planning better ways to navigate around the world, to adapt to it better, and to improve human life. Recognizing this, governments began to support scientific workshops and the scientists in them. The first scientific academies sprang up around this time. These included the Lincean Academy in Rome (1603), the Royal Academy in London (1660), and the Academy of Sciences in Paris (1666), all of them communities of scientists who helped each other and shared their findings. Governments recognized that their findings on things like food, medicine, housing, industry—and of course weaponry—were indispensable to many tasks of governing.

It was a terrific idea. What could possibly go wrong?

We'll see in Part II, when we meet three individuals who knew.

FOUR

GIAMBATTISTA VICO: GOING MAD RATIONALLY

CAN WE LOVE THE NUMBERS too much? Numbers, and the equations used to process them, are now used to guide almost anything—from creating symphonies and sitcoms to choosing colleges, careers, lovers, and pets. Calculations, metrics, and artificial intelligence have displaced ways people used to make decisions: relying on things like character, intuition, and moral compass. Politicians and strategists have taken these metrics to terrifying extremes. The algorithms used by Facebook and Amazon rely on them. On a more somber note, in *On Thermonuclear War* (1960), the physicist turned military strategist Hermann Kahn analyzed strategies for preparing, waging, and surviving nuclear combat. One of his publishers described it as "calm and compellingly reasonable," an attempt "to bring rationality to the public nuclear debate." Kahn wrote that we need to get past vague terms like "unthinkable" and "catastrophic," and instead approach thermonuclear war quantitatively. Even his opponents respected the book's rigor. The *Peace Catalogue* called it an excellent presentation of the "if-we-do-this-they'll-do-that school of strategic analysis." Kahn's book elevated calculating over living as a value.[1]

The eighteenth-century philosopher Giambattista Vico (1668–1744) grew up in the world that Bacon, Galileo, and Descartes sensed was coming. He realized how tempting but dangerous it was to use rational calculation as a universal tool. Some calculation is health-giving and stimulating, Vico argued, but too much is toxic. If scientific methods are taught to the exclusion of the humanities, it can cause individuals to "go mad rationally" and civilizations to slide into barbarism.

In his magnum opus, the *New Science* (published in 1725 under a title inspired by Bacon), Vico cast his argument in the form of a dramatic story. The story was no *New Atlantis*, painting a grand utopian vision, nor was it a *Discourse*, narrating a personal quest. The *New Science* was a story about how human powers developed from primitive to modern times—and then how they backslid. In this story, rational thinking is both a liberating tool and also, if allowed to displace the humanities, a corrosive one. Vico's tale identifies a potentially destructive side of workshop thinking that must be kept under control if it is to be beneficial to human society. Imagine if Bacon's island of Bensalem was originally populated by brutish humans, then slowly developed civilized society and established Salomon's House—but the very successes of Salomon's House then encouraged its citizens to become greedy and spoiled by all the benefits, and they declined into brutes again, with the cycle repeating itself.

THE MOST BRILLIANT DEFENSE OF THE HUMANITIES

Vico was another philosopher with a forceful and quirky personality. His own story is as dramatic as the one he tells in the *New Science*. "Vico's life and fate," wrote the philosopher and historian Isaiah Berlin, "is perhaps the best of all known examples of what is too often dismissed as a romantic fiction—the story of a man of original genius, born before his time, forced to struggle in poverty and illness, misunderstood and largely neglected in his lifetime and all but totally forgotten after his death."[2]

Giambattista Vico (1668-1744).

Vico was born in Naples in 1668, which, due to a massive influx of migrants from the countryside, was then the third-largest metropolitan area in Europe. The migrants included his father Antonio, who managed a small bookstall on the Via di San Biagio dei Librai, "Booksellers Row" in the heart of old Naples. Giambattista was the sixth of eight children, all of whom were crammed into a tiny apartment over the store.

As a child, Vico was determined and energetic, which helped him survive numerous challenges in his life. The first occurred when he was seven; probably fetching a book, he cracked his skull when he fell off a ladder. After he was unconscious for several hours, the doctor said the boy "would either die of it or grow up an idiot."[3] During the next three years, when he couldn't yet return to school, he read obsessively. His mother sometimes rose in the morning to find her son still awake, having read through the night.

Vico's reading was extensive, and by the age of sixteen he knew enough to successfully defend his father against a lawsuit. His success won him admirers, but similar to some others in this book he found practicing law distasteful. He preferred solitary study amid lots of books—which, by chance, he got. In a bookstore one day, he impressed a prominent bishop who hired Vico to tutor his nephews. They lived in a thousand-year-old castle in Vatolla, south of Naples, with stunning views of Capri. Vico moved to Vatolla in 1686, at the age of eighteen, and spent nearly nine years there. He had access to the well-stocked library of a nearby monastery, where he read extensively in philology—a term that covered what today is called the humanities—under a grove of olive trees. A library, time to read, a beautiful view: for a young, independent youth it was a dream come true.

Vico occasionally returned home to Naples during breaks from tutoring duties. In 1689, when he was twenty-one, he deferred to his father's wishes for him to become a lawyer and registered at the University of Naples. The university, founded in 1224, was the oldest state-supported institute of higher education in the world. Law was its most popular and prestigious field of study and a safe career direction, for as Vico's biographer H. P. Adams put it, "there were innumerable consolation prizes for mediocrity."[4] But it was not easy to become skilled. Italian law was a dense ragout of codes from different eras and regions stretching back centuries, so one had to master a wide range of ancient and modern sources.

Despite being thoroughly knowledgeable about the law, literature, and history of Naples and the region, Vico wrote much later, his solitary habits made him feel like an outsider in his native city, "not only a stranger but quite unknown."[5] He continued to register at the university and seems to have received a law degree around the year 1694. He resisted his father's entreaties to practice law, however, and continued to read on his own. He scraped together a living by tutoring and ghostwriting for affluent patrons, crafting letters, inscriptions, epitaphs, poems, and other writings, usually in Latin. He applied for university positions, but his independent streak and lack of mentors and family connections made it hard for him to get an entrée.

Finally, in 1697, he landed a job as the "royal professor of eloquence." It was a lofty title for a lowly position, what we'd now call an adjunct or lecturer. Yet it allowed him to move out of his father's apartment on Booksellers Row. He married a childhood friend who was also poor, they soon had three children, and Vico struggled to support the family. His financial struggles increased when one of his two daughters had costly medical problems and his son ended up jailed as a criminal. He continued to make ends meet by tutoring—for faculty, a forbidden but widespread practice—and ghostwriting, and would scrape by this way for the next forty-two years. All these duties left him next to no free time.

Fortunately, a university requirement forced Vico to write down his thoughts once a year. Each fall, at the university's opening ceremony, the royal professor of eloquence had to give an oration in Latin to incoming students and guests. Vico delivered his first in 1699.[6] But he was incapable of giving the usual boilerplate speech on how lucky the students were to be at the university, how proud their parents should be, and so forth. Instead, he argued that the students should engage in learning for its own sake rather than as a ticket for what he called "base material gain." These orations reveal him as a clever, thoroughly educated young scholar who defended the liberal arts and quoted extensively from philosophers, poets, and historians. His talks reflect a disorganized array of influences on his fertile mind.

Thanks to his readings, Vico had learned about the new scientific ideas propounded by Bacon, Descartes, and their followers. These ideas were highly controversial and had sparked a conflict in European capitals, especially Paris, concerning the best methods to promote education and knowledge-seeking. There were two warring camps with diametrically opposed positions: are the sciences really making us titans and masters over ourselves and nature, or are they trivial and distracting compared to the wisdom of the Ancients? The "Quarrel Between the Ancients and the Moderns," as it was called, raised serious issues: What is knowledge? What is wisdom? Is human progress possible?

Naples was a backwater, and its scholars were only beginning to learn about the quarrel. But they had to be careful picking sides. Spain, a conservative Catholic country, had ruled the city for two centuries, though it was embroiled in the War of Spanish Succession with Austria and several other European countries. The Spanish Inquisition still maintained tribunals in Naples who reviewed and censored books, and accused and imprisoned atheists. The Inquisition's officials labeled the "Modern" scholars "freethinkers" or "skeptics." These designations were only slightly less accusatory than "atheist," and intimated that people like Descartes and his followers could be condemned for questioning traditional beliefs and practices.

In his time at Vatolla, Vico learned to appreciate both perspectives. He was independent enough to appreciate the Modern methods, yet also to appreciate the value of the humanities that the Moderns were rejecting.

In 1707, in a defeat for Spain, Naples fell under Austrian control. When the university opened for the school year in 1708, its new managers wanted to stage the event with special pomp and pageantry. Vico boldly chose to discuss the conflict between the Ancients and the Moderns. He thought he could reconcile the two camps, describing a deeper conception of knowledge that would allow them peacefully to coexist. It was by far his most ambitious oration to date, and it led to his first book, *On the Study Methods of Our Time*. That book has been called "perhaps the most brilliant defense of the humanities ever written."[7]

ON THE STUDY METHODS OF OUR TIME

Vico revered Bacon, and praises him in the very first sentence of *On the Study Methods*. Bacon's *Advancement of Learning*, he wrote, sought to describe "what new arts and sciences should be added to those we already possess," seeking to "enlarge our stock of knowledge, so that human wisdom may be brought to complete perfection."[8] But while Bacon discovered a realm of new sciences, he was less interested in art, literature, and history as forms of knowledge and rational thinking. Bacon was more "the pioneer of a completely new universe than a prospector of this world of ours." The recent ascendance of the sciences raised the question of how best to teach and practice them. Descartes gave one answer, and his method was useful in narrowly defined contexts. But his influence had become so pervasive that his method was becoming widely regarded as *the* Modern way to study—carry out scholarly education and research—overshadowing and even replacing the old methods everywhere. To Vico, this was a mistake.

Vico then explained why.

Children can and do see the world differently than adults. Shaped by their immediate encounters with the world, children think imaginatively and metaphorically. This activity has its own "common sense," with its own knowledge and rationality. The common sense of children must be understood and nurtured if they are to grow into able citizens and intelligent thinkers. Adults need to nurture this by empathizing and playing along.

The Ancients cultivated children's imaginations in several ways, and Vico's description of these ways foreshadows the education he will later say is necessary to avoid going "mad rationally." One way is to cultivate children's natural bent for metaphorical thinking. To make a good metaphor requires insight, perceiving likenesses in apparently unrelated things, and pulling something new from darkness. Unlike Descartes, Vico saw clarity not as the beginning of the thought process, but as its end. Good metaphor-making is in turn a sign of what the Ancients called *ingenium*, or the skill of connecting disparate and diverse things. This word has no simple translation into English: the closest possibilities are "ingenuity," "wit" (as in "having the wit to do it"), "invention," and "engineering." Yet another study method of the Ancients was *ars topica*, even harder to translate, which refers to the process of filling the mind with knowledge so you can practice *ingenium* and teach it to others. "Thus," Vico writes, "without doing violence to nature, but gradually and gently and in step with the mental capacities of their age, the Ancients nurtured the reasoning powers of their young men."

The Modern study methods are different. Dazzled by Descartes's geometrical method, Moderns start by abstracting—reducing statements about the world to their simplest and clearest form—and then recombining them so they all interlock with certainty. This establishes formal chains of reasoning that proceed systematically and logically from one clear insight to another without passion or insight. The contrast between the Ancient and Modern study methods, according to Vico, involved imagination versus abstraction, metaphor versus demonstration, *ingenium* versus logic, and ambiguity versus certainty. Modern methods are wildly successful in the sciences. But their success deceived the Moderns into thinking they can be universally applied, even to areas where Descartes had not. Descartes had absorbed a lot of the humanities at La Flèche prior to propounding his method, but restricted that method to the sciences. His followers did not, and seized on it as a natural beginning point for all education and inquiry.

Vico found this educationally destructive. The geometrical method focused on thoughts that can be reduced to simple and abstract form, and insisted that these thoughts be connected without any suspicion of error. This has the effect of placing history, philosophy, language, all subjects taught in the *ars topica*, in the same category as knowledge that is false or uncertain—things to be put on mental hold or rid from our minds entirely. "Young minds," Vico wrote, "are too immature, too unsure, to derive benefit from it." The Cartesian method will stifle their common sense. It "may lead to an abnormal growth of abstract intellectualism, and render young people unfit for the practice of eloquence." Eloquence requires developing the relationship between speaker and audience, and appealing not "to the rational part of our nature but almost entirely to our passions." He continued, "The skillful orator, instead, omits things that are well known, and while impressing on his hearers secondary truth, he tacitly reminds them of the primal points he has left out and while he carries through his argument, his listeners are made to feel that they are completing it themselves."

Dazzled by the Modern methods, Vico wrote, "we pay an excessive amount of attention to the natural sciences and not enough to ethics." If this continues, our modern youth will grow up "unable to engage in the life of the community, to conduct themselves with sufficient wisdom and prudence." We have to make sure that our youth are taught "the totality of sciences and arts," nurturing their imaginations, memory, and common sense before introducing the Modern method. Before we become pioneers, looking for where we might go, we need to become prospectors, finding out where we already stand.

On the Study Methods is itself eloquent, a vibrant tribute to the Ancient methods even as it recognizes the (limited) value of the Moderns. Vico's language is full of vivid metaphors—such as prospectors versus pioneers—that exploit similarities between different areas of human experience. It does not analyze a problem but speaks with passion to an audience of students, parents, and dignitaries who periodically need to be reminded of the value of the humanities. Vico knows where his audience stands and how to move them to a new place. His book does what it says needs to be done.

It has some strange passages, however. Vico compares the enthusiasm of a scientist figuring out an equation to the Dionysian frenzy of a priestess being raped by a god. The French developed the Cartesian method, he writes, because their language lacks "great sublimity or splendor" and "is not fit for stately prose, nor for sublime verse"; its chief virtue is that it can "condense into a small compass the essentials of things." Vico writes, "We Italians, instead, are endowed with a language which constantly evokes images," which is better suited for common sense, metaphor-making, and mastery of the *ars topica*.

Vico was also disingenuous. He says he is simply refereeing the

Ancient and Modern study methods. He is not. He had deeply rethought them by proposing that "common sense," an imagination-rich connection to the world, is the foundation of all knowledge. Common sense, he thought, is not constructed of logic but weaves together things like memory, metaphors, and myths. Humans aren't calculating machines. People inhabit a space opened up by common sense that precedes the logical links that the Moderns celebrate, and that has to be cultivated before Cartesian-style rationality can be effective. No Cartesian "criticism" can cultivate that space.

Buried in the book, Vico says almost in passing: "we are able to demonstrate geometrical propositions because we create them." He spent the rest of his career exploring implications of this. The next year, 1710, between classes, tutoring, and ghostwriting, he found time to write another book that made this thought the centerpiece of an analysis of why the geometrical method is the wrong approach to the humanities. It was Vico's first analytical piece of writing, not composed as a speech or commissioned by a patron. He called it *On the Most Ancient Wisdom of the Italians, Unearthed from the Origins of the Latin Language*, a title inspired by Bacon's *On the Wisdom of the Ancients*. Again, Vico was disingenuous. He was not simply doing etymology—studying the origins and meanings of Latin words. His thought was new and original.

GOING MAD RATIONALLY

Vico began Chapter One of *On the Most Ancient Wisdom* with the sentence, "For the Latins, the true (*verum*) and what is made (*factum*) are the same thing."[9]

Take geometry. Its objects of knowledge—points, lines, planes—are products of our minds, made by abstracting and defining. So are its demonstrations, our shuffling around of these mental creations. We can therefore attain truth, and *certain* knowledge, in geometry because we make everything about it. We are insiders to its truths, so to speak.

After math, humans have two other sciences, those that study the forms of nature created by God (the natural world), and those that study human creations (the human world). We are outsiders to the first; only its Creator has the certain knowledge of nature that we humans have of geometry. We can observe nature's forms and see its patterns, but it's like trying to figure out chess by watching a game, without being able to see into the players' heads. But in the human realm we are insiders. We know what it's like to be the players and think like them; we have some idea of what they are trying to do. The human realm—traditional subjects of the humanities such as the arts, language, myths, political and social institutions, and their history—is composed of things humans themselves have created. We can know them, not by the geometrical method, but by what Vico called the "new art of criticism."

Vico thought that using Modern methods for the humanities was wrong for two reasons. The first is that common sense guides most everyday life activities in a way that cannot be quantified; if you applied the geometrical method to practical life, it would mean "going mad rationally." What if you tried to use Descartes's method to plan a speech? You'd never say anything clever, but would restate the obvious, patronize the audience, and come off as a geek (in the bad sense). The second is that common sense varies from culture to culture, so that without some interpretation you would never be able to appreciate the minds of people from different times and places.

So why does Descartes's method create such a buzz? Because it seems a shortcut to the truth. People think they can absorb knowledge in effortless little steps without the need for extensive readings of literature and for learning languages. You can't, Vico wrote. If our students knew as much of Plato, Aristotle, Epicurus, Augustine, and other Ancients as Descartes did, "then the world will have philosophers of equal worth."

We owe Descartes a lot, Vico wrote one critic, for demolishing our overconfidence in scholarly authority and for encouraging us to think for ourselves. Some of my best friends, he says, are "very learned Cartesians." But Descartes's method is too seductive. It relegates everything about which we cannot achieve geometric-like certitude to second-class status, including all aspects of civil wisdom contained in language, politics, eloquence, philosophy, and art. It only rearranges what we already know. Finally, it does not cultivate the skills necessary to find new knowledge.

Steeped in Renaissance humanism, with an appreciation for the vast productivity of the human mind, Vico spoke up for what the Moderns were leaving out of the educational curriculum. He sought to use the fact that humans create the humanities to give them new standing—even scientific standing.

Then over forty years old, Vico was still poor, surviving on a tiny salary plus tutoring and commissions. Another son—Gennaro—was born in 1715. Vico got a temporary financial boost in 1716 when he was paid ten times his salary for writing the military biography of a Neapolitan hero, but used most of it to finance the wedding of one of his daughters.

Then, in 1717, the occupant of a prestigious chair of law announced his retirement. The chair would not be filled until the retiree died, though the head of the university chose a certain Domenico Gentile as a temporary replacement, assuming his obvious incompetence would take him out of competition for the permanent job. The post seemed tailor-made for Vico. It paid six times his adjunct salary, which would solve his monetary problems and give him the prominent post he craved and deserved.

CAMPAIGNING

Vico spent the next five years campaigning for the position. He sought out leading scholars in Naples and abroad to give positive assessments of his writings for promotional purposes. He used his 1719 inaugural dissertation to lay out ambitious career goals. "Our present philosophy is uncertain, obscure, irrational, fabulous and quite incapable of being reduced to scientific principles," he told the crowd. He boldly set out to display all his knowledge of law and jurisprudence in a treatise published in Latin, *Diritto Universale* or "Universal Law," published in three volumes in 1720, 1721, and 1722.[10]

Universal Law tells the story of how Roman law evolved, from primitive times to Rome's becoming the greatest and most powerful city in the world with a legal system that exemplified formal justice. In the beginning, its inhabitants had to invent political and juridical institutions of their own. Initially, these were crude and reflected the interests of the powerful, but they slowly evolved via what we now call class conflict. Previous scholars, including Vico's colleagues at the University of Naples, treated law and justice either as a set of universal rational principles good for everyone and everywhere ("natural law"), or as a specific set of prescriptions by wise legislators ("law of nations"). Vico saw a creative third possibility: laws are historical products of human beings struggling to realize goals within specific social and institutional contexts.

Human beings are motivated by self-interest, but the historical context shapes what institutional changes they seek as well as the effects of those changes—and thus the character of the people who live amid the institutions. The result was a long struggle between patricians and plebeians. The effect of the struggle was to slowly transfer authority from powerful individuals to institutions that acknowledged the rights of the plebeians. In the process, new forms of justice emerged in marriage, trials, office-holding, citizenship, and other institutions. What justice and citizenship meant, Vico demonstrated, depended on what part of Roman history you were talking about.

THE NEW SCIENCE

The chair of law's retired occupant died in December 1722. When the competition for a successor was formally announced in January, Vico, now fifty-five years old, was the first to apply. Thirteen more candidates followed, including Gentile. Vico was sure he was a shoo-in, and should have been. But the jury's twenty-five members split almost evenly between those favoring the faculty and those more loyal to the city's rulers. The former rallied to a bland, uncontroversial scholar, the latter around the docile flunky Gentile.

Vico was oblivious to the academic politics. He stayed up until five in the morning on April 9, 1723, crafting an eloquent speech to deliver the next day as part of the hiring process, in which he cast his work as the cutting edge of a movement known as legal Humanism. But as the historian Barbara Ann Naddeo remarks, "his dazzling erudition was marred by his unrivaled presumption." She continues, "Not only had Vico presented himself as assuming the mantle of the legal Humanists, but he also had peppered his talk with Greek technical terms and expressions, which he apparently pronounced with some difficulty and resulted in at least one noteworthy moment of pregnant silence during his delivery." Vico's efforts were to no avail. Gentile won by a single vote over the candidate preferred by the faculty, 13–12.[11] Vico failed to get even one vote. "A sorry piece of political theatre," Naddeo concludes.[12]

Vico was so devastated that, in his autobiography, he cannot bring himself to admit that he withdrew at the last minute to spare himself humiliation. But the disaster was one of the best things ever to happen to him. Had he won, he probably would have spent the rest of his life churning out pedestrian essays on predictable subjects. Success would have made him a second-rater. Instead, Vico wrote, he gave up "all hope of ever holding a worthier position in his native city," threw caution to the winds, and set himself feverishly to craft a fresh way to present his principal idea. He set out to write a *Novum Organum* for the modern world, updating Bacon's classic work so that it included a fundamental role for the humanities. Vico wrote it in Italian, the language of his countrymen, and called it the *Scienza Nuova*, or *New Science*. It was his magnum opus. God bless Naples for handing me this setback, he wrote years later. "Could I owe it a greater debt than this?"[13]

Vico spent the rest of 1723 and all of 1724 on the project, completing a two-volume work in which he unveiled his ideas while critiquing his predecessors. It was so lengthy that he could not afford to publish it himself. In such cases authors often sought a patron to underwrite the expenses in return for the author dedicating the book to them. At the end of 1724, Vico found one: Cardinal Lorenzo Corsini.

But the road to the *New Science* was paved with more disasters. A few months later, in July 1725, Corsini sent Vico an apologetic letter saying he had overspent his budget and could no longer afford to sponsor Vico's book. Again he was devastated. A friend suggested trying what we now call crowdsourcing, lining up paying subscribers. Having lived among booksellers and academics, Vico knew this was hopeless. He therefore decided to pay for his book himself by selling his one valuable possession—a precious ring—that only brought enough to pay for a quarter of the pages. He then made the gritty and pragmatic decision to radically shorten the book by presenting his work not as a critical commentary on his predecessors but in positive form as his own view.

Rewriting it took him almost two months. He also changed the way the book was to be printed. Instead of the standard quarto format, where each printed sheet would be folded into four leaves for a total of eight pages each, he had it printed in small type—brevier, or about eight-point type—so that it could be crammed onto twelve sheets of duodecimo, or sheets folded into twelve leaves, making for a single volume of 288 pages. Vico dedicated it to Corsini anyway, still hoping for reimbursement, and sent him a copy; Corsini (soon to become Pope Clement XII) gave it to a friend, who never read it and left it to the Vatican library.[14]

Disaster was Vico's muse, poverty his editor. These broke his defenses, forcing him to abandon the verbose academic style of the day. Out jumped a new understanding of science and human history. The *New Science* uses what Vico calls "a new critical method for sifting the truth . . . from popular traditions." His biographer Adams writes, "The light falls no longer among rocks and chasms; it is diffused over a wide landscape. The change of language is also a great gain. The wintry dignity of Latin gives place to the sunny vitality of Italian."[15]

Vico published the first edition of the *New Science* at the end of September 1725. Letters he sent to friends that October with copies of the book show that he knew its power. My hardships have strengthened me, he wrote, and this book "has filled me with a certain heroic spirit, so that I am no longer troubled by any fear of death."[16]

The *New Science* draws on Vico's earlier works but recasts their implications and makes them available to a wider audience. Naddeo writes that, compared to *Universal Law*, which was long, in Latin, and full of tendentious critiques, the *New Science* is a "popularization." That's less of an exaggeration than it seems. While most of Vico's previous work was aimed at Neapolitan scholars, the *New Science* was grander, covered the full sweep of the humanities, and served as "a platform from which to make an appeal to the literati and scholars" across all Europe.

OUTLINE OF THE *NEW SCIENCE*

Still, it's no beach read. Vico often loses himself in details, and engages in flashbacks, digressions, and extraneous remarks. It's full of inconsistencies, improbable claims, pointless critiques of predecessors, and too much information about fine points of law. It is confusingly organized into Books, Sections, Chapters, Elements, and Principles with numbered paragraphs. Still, it's a grand story. Here is a brief outline.

1. *The human world is knowable because humans made it.* Even when we study primitive societies, these predecessors are not alien but belong to the same cultural evolutionary lineage that we do.[17]
2. *A new art of criticism is required to understand these human-made worlds.* This art is not like understanding mathematics, but more like understanding what it means to be afraid, proud, or in love. It uses the methods of the Ancients—imagination, *ingenium*, empathy, and metaphorical thinking—to understand language, myths, and institutions.
3. *This new art of criticism aims to grasp the common sense of the peoples it studies.*[18] Common sense is the rationality expressed throughout a nation's practices and institutions, embodied in how its people react to everything from thunder and lightning to justice and death. Common sense is not so much a "Theory of Everything" as a "Sense of Everything" (SOE). This SOE provides an open space in which human beings encounter, grasp, and respond to the world. It gives concrete shape to what humans do to realize their desires and ambitions.[19]
4. *Human actions have unanticipated consequences.* How humans act out of their SOE in seeking to satisfy their needs and desires has unanticipated consequences that, in turn, change their SOE. The result is a progressive development of human institutions.[20]
5. *These unanticipated consequences take human history through three stages: poetic, heroic, and rational.* In primitive times, the SOE was poetic, dominated by imaginative thinking; the consequences of people's actions slowly transformed their SOE into phases where invention and then reason prevailed.[21] Vico did not

romanticize primitive humans, having no illusions about how harsh and brutal plebeian life was. But he thought it important to study them as the beginning of the organic process by which humanity evolves into an age where reason dominates.[22]

6. *Humans thereby make their own history*. History does not unfold by chance (as per the Epicureans), nor is it predetermined (Stoics), nor is it due to God's interventions (theologians), nor to great and wise individuals (historians). History is the outcome of humans trying to achieve their desires and ambitions, working in the context they were born into, with their actions changing that context, pushing humanity further.[23]
7. *An Ideal Eternal History lies behind these three stages*. Rome is not merely an interesting case study of how humans develop ideas of law and justice but a paradigmatic case of an ideal pattern.[24] This pattern provides a maturational yardstick for human civilizations. In describing this ideal pattern, the *New Science* becomes a science rather than a history of human ideas and interpretations. Thinking is finite and local, always inheriting and transforming, but it follows a discernible path. To find this path, the new scientist has to piece together the string of causes by which humans make history. Only then can we *know* the human mind in the sense that we know what we make.
8. *In grasping this Ideal Eternal History, the New Scientist achieves human self-knowledge*. In the "I think" of Descartes, the "I" takes too much credit. Everything—the entire human world and its history—is involved in that thinking. Awareness of our existence doesn't necessarily mean self-knowledge. Why I am *this* particular person speaking *this* particular language and following *these* particular customs is not something I intuit but inherit.

9. *Human history contains the seeds of its own slide back into barbarism*. The story does not have a happy ending. At just the time humanity reaches the age of reason, it begins to slide backwards. Humans fall "into the custom of each man thinking only of his own private interests" in a "barbarism of reflection." The maturation of thought fosters an overreliance on calculation and analytical rationality and the atrophy of imagination, ingenuity, and the humanistic arts, leaving humans to think that they are mastering the world at the same time as they are losing it. Human history evolves, but can also devolve.

THE BARBARISM OF REFLECTION

The barbarism of reflection is one of Vico's most daring, insightful, and original thoughts. Many stories about Rome's downfall, especially those told by Christians like St. Augustine, blamed moral causes. Rome, according to these tales, declined because its citizens lapsed into gluttony, sensuality, and self-interest. This allowed the government to be taken over by demagogues, crowd pleasers, and others with little insight into governing. Vico identified a different, surprising, and deeper, cause: the success of rationality itself.

Bacon's *New Atlantis* was a sunny and optimistic story: when humanity turns to science, it liberates itself and creates a new Garden of Eden in which the benefits of science are obvious to and welcomed by all. He had not seen any significant dark sides to science. Vico lived in a world that Bacon helped to make possible, and saw dangers that Bacon did not. Reflection, so encouraged by the Modern methods, involves the ability to reject tradition and act only for self-interest. Paradoxically, the very force that humanity finds so liberating—science and reason—if overused makes social bonds dissolve and alienates humans from each other. Just as Modern educational methods can lead individuals to go mad rationally, on a social level they lead to a slide back into individualism and egotism. European history, like that of the Roman Empire, planted the seeds of its destruction in the success of its sciences.

Can humanity's slide back into barbarism be stopped? No god is rooting for humanity behind the scenes, Vico thought. No politician has the power to pull it through. No Gaia or planet-spirit is there to step in and save humanity from itself. Humans themselves are both the cause of their own cultural disease and the only conceivable source of the remedy. The slide could only be halted, Vico implied, if a shock to the system resulted in a wholesale reworking of how the humanities were incorporated into human life so that humanity took itself out of this historical tractor beam. Vico thought the *New Science* might have this impact.[25]

Vico spent the rest of his life reworking the *New Science* into two more editions, the third not finished when, mentally and physically debilitated, he passed away in 1744. Even after his death, fate seemed to have it in for him. On the day of his funeral, two groups of colleagues showed up to accompany his body to the church: the university professors and a group of priests. They argued over which group would carry the casket, and the priests, in a huff, walked away. The professors then realized they could not conduct the service without the priests and left. Vico's devoted son Gennaro, who succeeded him in his professorship, had to carry his father's corpse back to the house and arrange the funeral for another day. Vico was buried in the church of the Gerolimini, a block away from his father's bookstore on the other side of the university.

Vico's work took time to have impact. Part of it was his fault: he could not resist adding material to the third edition of the *New Science*, making it longer and more difficult to read. Also, his lowly position meant that his work never achieved the visibility that the work of a more prominently placed scholar would have had. As his biographer writes, "It was as if a great ship had been built, capable of navigating all the oceans of the world, and was left moored in the dock of the shipbuilder to be visited occasionally by a few friends of the inventor, and mentioned in their correspondence by one or two superior persons who recognized not so much its value as the cleverness that must have gone into its construction."[26]

Yet Vico's circle of appreciators slowly grew. His work has recently attracted more notice, among other reasons because it was so prescient about what fuels science denial. The Modern methods can indeed tempt humans to love the numbers too much. If these methods are overused, they can inhibit rather than promote the flourishing of the human world. Science deniers often exploit this vulnerability. When scientific findings run against a policy—any policy—that deniers like, they can suggest that extending scientific methods to that area is overreaching.

In arguments over *Gill v. Whitford* (2017), to cite a recent instance, US Supreme Court Chief Justice John G. Roberts Jr. made just that suggestion. The case concerned partisan gerrymandering, or the drawing of legislative maps to favor one party over another. Plaintiffs had argued that Republican officials in Wisconsin had drawn the state's legislative map in such an extremely partisan way that it unconstitutionally denied representation to certain kinds of voters. The plaintiff's lawyers had used mathematical methods to quantify the partisanship of the gerrymandering. Chief Justice Roberts, a Republican, would have none of it. During the oral arguments, he said that using mathematics means "taking these issues away from democracy," and rejected it as "sociological gobbledygook."[27]

Vico's notion of the barbarism of reflection also diagnoses another motive for science denial. The very success of science and technology has encouraged the illusion that almost everything is within our grasp. We are accustomed to relying on science and technology but resent having to pay for it. As one US congressman remarked, "Why do we need Landsat satellites when we have Google Earth? Why do we need weather satellites when we have the Weather Channel?" Modern humans feel like free agents, entitled to choose their forms of energy, nutrition, and environmental conditions without having to make severe, costly, and risky trade-offs, an illusion that's amplified by ideology, political influence, and money.

Vico was the first to see clearly that the very success of the sciences nurtures the grounds for their rejection by outsiders. Had Vico ever written a *New Atlantis*, it would have supplemented Bacon's by combining scientific methods with humanities education—and having all citizens, or at least the leaders, aware of the story of what happens when you don't. Otherwise, this New Atlantis would be doomed to suffer a fate similar to its predecessor's.

FIVE

MARY SHELLEY'S HIDEOUS IDEA

"How," Mary Godwin Shelley asked herself, did I come up with "so very hideous an idea?"

It was summer 1816. She was visiting Switzerland with her lover, the poet Percy Shelley, and the couple was staying near the poet Lord Byron and his companion John Polidori. But it rained almost every day and remained chilly, so instead of hiking they spent days inside Byron's villa reading ghost stories. They read all they could get their hands on. Byron then suggested they each write their own. The men set to work. Mary wanted to find something truly scary—something to awaken the "mysterious fears of our nature"—but instead got writer's block. That was mortifying, as she was the child of two famous writers and the partner of another. Every morning the men would ask if she had thought of anything. "Nothing," she'd reply.[1]

One day she recalled a conversation between Byron and Percy about the principle of life. They mentioned scientists who had used electricity—then a new technology—to make parts of dissected animals flinch. She couldn't sleep that night and had wild dreams. She imagined a technician sending a spark into an assemblage of body parts. The thing began to move. She imagined the technician, terrified, praying for the movement to stop—and running away, terrorized, when the creature kept going. She trembled as she pictured the newly born, now immortal thing opening its "yellow, watery" eyes.

The next morning, Mary told her companions that she had an idea.

THE BACKGROUND

Mary Godwin Shelley (1797–1851) tells the above story in the introduction to the second edition of *Frankenstein* (1831). She was not the only writer who seems to have spun the origin story of her most famous work to suit her own purposes. Even more spinning of the story, though, has taken place in two centuries of theater and film adaptations. As a result, even those who have not read the book know a canned version of what seems the basic plot: an out-of-control monster built by a well-meaning but careless scientist named Victor Frankenstein.

That canned version is even embedded in contemporary everyday language. The prefix "Franken-," as in Frankenfoods, Frankenfish, and Frankenpets, is often used to create a charged word referring to something terrible that should not have been created because it is unnatural and the product of unrestrained consumerism. Frankenwords are catchy, wave red flags, and are designed to provoke fear. They seem to endow those who use them with unclouded and superior moral judgment.

But the story that unfolds in Shelley's novel is not so simple, nor is the moral clear. The story has to be read carefully, against the grain of the theater and film adaptations that have made it so familiar. Like the monster itself, the basic elements of the novel were already around at the time that Mary Shelley wrote. She provided the spark that fused them together—and after the novel sprang to life, it has haunted us ever since. It is still a lightning rod for the anxieties of living in a techno-scientific age. Today we think we find the monster in many guises: genetically modified foods and organisms, clones, radioactive sources, climate-altering chemicals, omnipresent drones, and nanoparticles, to name a few—whose long-term effects are not completely known and may not pass away as conveniently as Shelley's monster.

Mary Godwin Shelley (1797-1851).

Shelley's inner circle had many scandals and entanglements whose effects on her may also have helped give birth to Frankenstein. These entanglements—which include sometimes incestuous affairs, pregnancies, births, miscarriages, and deaths—all figure in the backstory. Mary Godwin Shelley was the daughter of William Godwin, a utopian thinker and educator, and Mary Wollstonecraft, a pioneering feminist author. Wollstonecraft died of complications from childbirth shortly after her daughter was born. Godwin remarried a woman with two daughters, Fanny Imlay and Claire Clairmont. Thanks to Godwin's tutoring and support, Mary was educated in politics, philosophy, and history.

She was bolder and rather more imperious than even her utopian feminist progressive father liked. When Mary was sixteen, she met Godwin's political disciple, the poet Percy Bysshe Shelley. The two met secretly at her mother's grave. Godwin was furious to find out that his daughter was involved with a married man, and revoked his previous judgment of marriage being an oppressive institution. His radicalism covered politics and education but not parenting. His headstrong daughter ran off to France and Switzerland with Shelley, who left his pregnant wife Harriet behind. The two lovers were accompanied by Mary's younger stepsister Claire, now sixteen years old and her rival for Shelley's affections.

The trio returned to London in the fall of 1814 with Mary pregnant. Percy and Mary stayed together despite debt, social ostracism, and a string of stressful events. These included Harriet's giving birth to Percy's son Charles that November. A few months later, Mary gave birth prematurely, and the child died. But her father, still furious, refused to see them. Mary was soon pregnant with another child, William, who was born at the beginning of 1816.

Late that spring Mary and Percy took off for Geneva, posing as husband and wife for hotel registration purposes. Again Claire accompanied them. Mary's younger stepsister, adventurous herself, now had her sights on Lord George Byron, with whom she had recently had an affair in London and by whom she was now pregnant. Claire had learned that Byron, too, was headed for Geneva, and she intended to continue pursuing him (successfully, in the end) there.

Four years older than Percy Shelley, Byron had become a celebrity in 1812 at the age of twenty-four, following the publication of two cantos of a poem called *Childe Harold's Pilgrimage*. "I awoke one morning and found myself famous," he later wrote.[2] Byron, too, was a scandal magnet. His numerous real and purported affairs, aside from Claire, included his half sister, whose child he was rumored to have fathered. In January 1815, Byron had married Baroness Annabella Wentworth, and that December their daughter Ada was born. Ada was Byron's only child not born out of wedlock. But by January, Annabella had had enough of Byron and left. A few months later, beleaguered by more scandal, minus an heiress and fully in debt, Byron fled London for Geneva, where he intended to complete a third canto of *Childe Harold*.

Accompanying Byron was John Polidori, a twenty-one-year-old physician who wanted to become a writer. Charming and handsome, he had convinced Byron to bring him along as his personal physician. Byron did not seem to mind that Polidori exploited their relationship for personal gain. Before they departed England, a publisher paid Polidori to keep a secret diary of his trip with the famous poet for publication. The publisher must have been disappointed, for much of the diary was about the crush Polidori soon developed on Mary.

At a Geneva hotel, Claire reunited Mary and Percy with Byron and Polidori. The foursome decided to move out of the Geneva hotel and rent houses near the shores of Lake Geneva. Byron and Polidori moved into an expensive country villa called Belrive, which Byron rechristened Villa Diodati after its owner. Mary and Percy, and a nanny to take care of little William, moved to a smaller house nearby.

The Villa Diodati had a garden, balcony, and spectacular view of the lake and the Jura Mountains. It still exists and is a few minutes' walk outside Geneva. It is privately owned, and a wall shields most of the house from the street. A plaque is positioned on the house just above the wall, visible to passersby, which mentions Byron's famous narrative poem.

LORD BYRON
Poête Anglais
auteur du
PRISONER of CHILLON
habita la
VILLA DIODATI
en 1816
y composa le 3me chant
de
CHILDE HAROLD

The foursome spent a lot of time that summer in the Villa Diodati, thanks to a climate-changing event. The previous year, one of the most powerful volcanoes in human history had erupted in Indonesia. Tambora, as it was called, poured enough ash and debris into the atmosphere to change weather all over the globe. Europe, some 7,000 miles away, experienced clouded skies, unusually low temperatures, and record amounts of rain. It was, Mary wrote, a "wet, ungenial summer, and incessant rain often confined us for days to the house." The foursome took trips when the weather broke, including one to the Mer de Glace on the side of Mont Blanc, one of the glaciers most accessible to European tourists. In Mary's novel, this glacier is the site of the monster's first confrontation with his maker. But mostly the four writers spent time inside consuming what books the villa offered them.

THE GENRE

These books included several gothic horror tales. A typical plot revolved around a past sin of the protagonist—betrayal, adultery, murder—that resurfaces, supernaturally, to destroy that person's future. Mary describes two she read that summer which remained "as fresh in my mind as if I had read them yesterday." In one, a man who had abandoned his bride returns remorsefully to her, but whenever he tries to embrace her, he finds himself in the arms of a ghost. In another, a man who has strayed from familial duties kisses his children as they sleep, only to find that his kisses kill them.

Frankenstein, the novel Mary would write that summer, combined the gothic horror plot with elements of what we call science fiction. Science fiction plots often revolve around human-made innovations that have reshaped the world. Stories of statues coming to life, and of humans creating humanoid forms, have been around since ancient Greek times. In *Frankenstein*, Mary dropped that kind of creation into the very spot occupied in gothic horror fiction by a sin.

Another of *Frankenstein*'s innovative elements was that the unleashed horror was potentially universal; it threatened all human life and was not just limited to the sinner and those closest to him or her. The monster nearly coerces Frankenstein into creating a mate for his monster, with the possible outcome being "a race of devils" to wander the Earth "who might make the very existence of the species of man a condition precarious and full of terror."

THE TECHNOLOGY

Electricity was featured in the popular science of the late eighteenth and early nineteenth centuries. In the 1750s, well-publicized experiments with lightning by Dalibard in France, Franklin in the United States, and Lomonosov in Russia had acquainted the public with electricity's marvels and awe-inspiring power. Electricity was a spectacular thing that broke down the distinction between the natural and the artificial. Working with it was no longer simply controlling some material or process in nature—like fire, water, or steam—as if science were some expansion of an arts and crafts activity. Electricity was a phenomenon that was somehow "in" nature, which popped up here and there in apparently unrelated forms (static electricity, shocks delivered by some eels, lightning), but which humans had to prepare first before it could be used at all. Electricity helped to change the popular understanding of nature, as well as of science.

Electricity's wonders continued to grow in the 1780s, when the Italian physician Luigi Galvani discovered that electricity can make muscles of dead frogs twitch, a phenomenon called galvanism. The invention of the battery in 1800 by the Italian physicist Alessandro Volta further inspired electrical research. Galvani's work became well known in England when, in 1803, his nephew Giovanni Aldini gave a demonstration in front of royalty in which he used a Volta battery to electrically stimulate the head of an ox, causing its eyes to open and its nostrils to swell. The following year, Aldini electrically stimulated the body of an executed criminal, causing the corpse's jaws to quiver, its muscles to clench, and one of its eyes to open.

Mary was more acquainted with electricity than most of the general public thanks to her father and to Percy. William Godwin was a personal friend of Humphrey Davy, an English chemist who worked with electricity, while Percy had become fascinated by electricity in boarding school. As a prank, he would use electrical equipment to charge his doorknob with static electricity to shock visitors. While at Oxford, Percy's companion Thomas Jefferson Hogg (another of Mary Godwin's lovers) recalled Percy charging himself up to make "fierce, crackling sparks" fly forth, so that "his long, wild locks bristled and stood on end."[3]

In 1814, after Mary and Percy returned from their first trip to Europe, they lived near an amateur scientist named Andrew Crosse, who had built an electrical laboratory in his house in which he would give demonstrations. That same year, Mary and Percy followed a heated public debate at the Royal College of Surgeons between one doctor who held that the body was simply a machine and another who maintained that there had to be some "principle of life" that might be linked with electricity.

"Mary Shelley based Victor Frankenstein's attempt to create a new species from dead organic matter through the use of chemistry and electricity on the most advanced scientific research of the early nineteenth century," Shelley's biographer Anne K. Mellor has written. "Her vision of the isolated scientist discovering the secret of life is no mere fantasy but a plausible prediction of what science might accomplish."[4]

THE TRIGGER

Frankenstein was born during an emotionally charged vacation, amid scientifically intense discussions and an environmentally stormy climate, when Byron kicked off the ghost story writing venture one night in the Villa Diodati.

Percy wrote a poem called "A Fragment of a Ghost Story." Byron wrote "Fragment of a Novel," an incomplete vampire story that his publisher included, without Byron's consent, as an appendix to a book containing one of Byron's poems. Polidori began something that didn't pan out, but later reworked Byron's fragment into a novel called *The Vampyre*. That was published under Byron's name and became an inspiration for Bram Stoker's famous story *Dracula*.

Mary started writing her story in Geneva, and then worked on it after she and Percy returned to London in September 1816. The crises and entanglements of their extended circle intensified over the next few months, and Shelley worked on it whenever possible. In October 1816, Mary's older half sister Fanny committed suicide at the age of twenty-two with an overdose of laudanum, an opium-containing substance. Harriet Percy, Shelley's wife, took another lover, became pregnant, and in December 1816 herself committed suicide by drowning. Three weeks later, on December 30, 1816, Percy and Mary Godwin (now pregnant again) were married. Her father William, ending his estrangement from his daughter, attended the wedding. Mary's stepsister Claire gave birth to a daughter, Alba (Allegra), fathered by Byron, two weeks later in January 1817; Byron was in Italy, never to return to England. Mary gave birth to her second child, Clara, in May. Three children—Alexander, Allegra, and Clara—were baptized together in the same ceremony the next year.

Mary Shelley finished her manuscript in May 1817, and the book appeared in three volumes in January 1818. Publishing books in three parts was common in the nineteenth century. Paper was expensive, and circulating libraries were a huge part of the book-buying market. Three-part books made more money, for they tripled the number of subscribers for a book. Mary dedicated *Frankenstein* to her father. It was published anonymously, without a name on the title page, and reviewers (including Sir Walter Scott) assumed a man had written it. "There never was a wilder story imagined," wrote the *Edinburgh Magazine*; "yet, like most fictions of this age, it has an air of reality attached to it by being connected with the favorite projects and passions of the times."

Mary Shelley's personal life was strange and heartbreaking. All the men who had been at the Villa in the summer of 1816 were dead within two years of *Frankenstein*'s publication. At the end of 1818, the Shelleys left London with their two children William and Clara, and took Allegra to deliver to Byron in Italy. While the couple was in Italy in 1818 and 1819, both of their own children died. In 1821, Polidori, saddled by depression and gambling debts, committed suicide. Allegra died of fever at age five in 1822. Two months later, Percy Shelley died while on a boat that capsized in a storm. In 1824, Byron died of poor medical attention while preparing to join the Greek War of Independence.

Mary and Claire lived the longest. Mary, who wrote several novels and memoirs, died in 1851 at the age of fifty-three. Claire outlived nearly all the members of the Shelleys' circle, dying in 1879 at the age of eighty. Ada, Byron's daughter with Annabella and his only non-out-of-wedlock child, became a mathematician and worked with Charles Babbage, inventor of the computer. Ada Byron Lovelace is credited with being the first computer programmer.

THE STORY

Frankenstein consists of several stories nested within each other. A man named Walton leaves his devoted sister for the North Pole. It is a quest he has dreamed of since childhood, but which his familial duties have prevented him from pursuing. Walton assumes that the North Pole's purity, remoteness, spectacular views of the heavens, and accessibility to the power of the Earth's magnetic source will enable him to discover new knowledge to help humanity. But the journey is cut short when his vessel becomes icebound.

A frozen, exhausted man appears on a dogsled with only one dog left alive. He faints after being brought onboard, but brandy revives him. When the two men speak, Walton is the first to tell his story—but the story makes the other man, Victor Frankenstein, livid. I, too, was on a quest to seek "knowledge and wisdom" and to conquer "the ever-varied powers of nature," Victor says, telling Walton to give it up. Frankenstein then tells his story. This set-up—one knowledge-obsessed man telling another his story—allows for a lot of moralizing.

Victor was from a distinguished family from Geneva, and lived with an adopted sister, Elizabeth, and a younger brother, William. Their house is named Belrive. While Elizabeth was the "living spirit of love" and William sweet and sensitive, Victor was rebellious. His siblings enjoyed the family's happy domestic life, Victor tells Walton, but "the world was to me a secret which I desired to divine." He threw himself into reading metaphysics. One day when he was fifteen, he watched lightning shatter an "old and beautiful oak" into "a blasted stump." Out went Victor's interest in metaphysics, which he now regarded as "a deformed and abortive creation" and "not real science." He was now fascinated by electricity and galvanism. He began to pursue mathematics as an entrée into science "built upon secure foundations."

Just before Victor left for the university in Ingolstadt, Germany, Elizabeth contracted scarlet fever. While nursing her back to health, Victor's mother contracted it as well and died. The tragedy left Victor with a "void" in his soul.

At Ingolstadt, Victor pursued his new enthusiasm for science. A professor tells him that modern scientists perform miracles. "They penetrate into the recesses of nature, and show how she works in her hiding places," and they "have acquired new and almost unlimited powers; they can command the thunders of heaven, mimic the earthquake, and even mock the invisible world with its own shadows."

Victor was hooked. "I will pioneer a new way," he promised himself, "explore unknown powers, and unfold to the world the deepest mysteries of creation." He threw himself into his studies. "None but those who have experienced them can conceive of the enticements of science." In other fields, people have been there before, and only geniuses make discoveries—but in science even those of modest intellects can contribute. Victor became particularly interested in the "principle of life," and began to study it by dissecting corpses.

Here, and often in his story, Victor pauses to lecture Walton on the dangers of seeking knowledge and how much happier people are who make their hometowns their worlds. Reading quickly, one might assume that the "danger" is the unleashing of mysterious forces. But Victor also means that knowledge itself is a source of unhappiness, because those who seek it cannot go home again or even rest content, for each discovery brings new questions. Victor is also warning against the hubris of trying to control nature through science.

His moralizing over, Victor returned to his tale. In the quest to create life, he spent two years collecting body parts and building electrical equipment. One rain-spattered November night, he succeeded. "I saw the dull yellow eye of the creature open; it breathed hard, and a convulsive motion agitated its limbs." Victor had animated an eight-foot-tall creature with horrible watery eyes, a shriveled complexion, and straight black lips. When he saw it, "the beauty of the dream vanished, and breathless horror and disgust filled my heart." He raced out of the lab, paced his bedroom, and after falling asleep had nightmares of embracing Elizabeth, only to have her wither away and die. After Victor awoke, he looked in the laboratory and saw the monster staring at him. Terrified, he fled outside and encountered his dearest childhood friend Henry, who had worried about him and unexpectedly come to visit. When Victor returned home, he contracted a fever that felled him for months. Meanwhile, the creature roamed the countryside by itself.

When Victor eventually returned to Geneva, he discovered that his brother William had been murdered in a park called Plainpalais. The murder was blamed on a trusted servant, who was tried, convicted, and executed. But Victor realized that the monster was the culprit, making Victor himself the murderer "not in deed, but in effect." Now indirectly responsible for two deaths, Victor lived in guilt and fear.

Statue of Frankenstein's monster at Plainpalais, Geneva.

For a while, he sought solace in nature: "the unstained snowy mountain-top, the glittering pinnacle, the pine woods, and ragged bare ravine; the eagle, soaring amidst the clouds—they all gathered round me, and bade me be at peace." He visited the Mer de Glace, which in a previous visit had filled him with "sublime ecstasy," transporting his soul "to light and joy" and causing him to forget his woes. "The surface is very uneven, rising like the waves of a troubled sea, descending low, and interspersed by rifts that sink deep. The field of ice is almost a league in width, but I spent nearly two hours in crossing it."

On the glacier, he saw the monster clambering over the ice to reach him. Creator and creation confronted each other directly for the first time. Over the next six chapters, the heart of the book, the monster tells his tale—another story within a story. The condensed plot of this story is that, while the creature began with love and respect for humans, he encountered from them nothing but hatred and repeated attempts to kill him. At the end of his tale, the creature mentions reading *Paradise Lost*; he regards himself as an Adam without a God to care for and guide him, leaving him "wretched, helpless, and alone." He demands that Victor create a female for him so that he might experience the "affections of a sensitive being." Remarkably, while Victor Frankenstein comes across as pathetic and obtuse, his artificial creature comes off as sensitive and deeply human, even and especially in his rage and destructiveness.

Victor agrees to create a female monster, but ultimately changes his mind and destroys it. In the horrifying end of this part of the story, the monster takes revenge by killing Victor's closest friend Henry as well as Elizabeth—on the very day she marries Victor.

The narrative then shifts back to Walton's story. On the icebound boat, Walton tends to the dying Victor. In a final bit of moralizing, Victor warns Walton to "avoid ambition," even to shun "science and discoveries." The monster appears, kills Frankenstein, and at the end of the novel heads off across the ice to burn himself on a funeral pyre.

OUT OF CONTROL

In 2014, two humanities scholars at the University of New Mexico published an article in *Science and Engineering Ethics* that envisioned Victor Frankenstein submitting his research to an institutional review board (IRB). IRBs are panels of the sort now mandatory in the United States for preevaluation of research involving human or animal subjects, and they are required to follow strict procedures. "Had Victor Frankenstein had to submit an IRB proposal," wrote the authors, "tragedy may have been averted, for he would have been compelled to consider the consequences of his experiment and acknowledge, if not fulfill, his concomitant responsibilities to the creature that he abandoned and left to fend for itself."[5]

The article cleverly exhibits *Frankenstein*'s value for teaching contemporary research ethics. At the same time, though, the article exemplifies the familiar but erroneous way in which the story is understood: as a tale about the creator and the creation. To determine the ethics of the experiment, the experimenters followed standard procedures and based their judgment on whether Frankenstein acted carelessly or unethically, and on whether his project had the potential to do harm. In this approach, blame is placed either on the creators—the scientists who make the bombs, greedy CEOs of Big Pharma, unfeeling doctors, and so forth—or the creations—nuclear transformations, genetic manipulations, ecological alterations, and so on.

Numerous film and theater adaptations of the Frankenstein story reflect this approach. In the famous 1931 film *Frankenstein*, starring Boris Karloff, the creator unwittingly installs an abnormal brain in the creature. The creation, a property destroyer and serial killer, is an evil criminal from the start. Other readings of the novel blame Victor, the monster's creator. At one point in Shelley's story, he calls the monster "my own spirit let loose." In these readings, Victor is motivated by internal demons, such as the trauma of losing his mother before he leaves for the university. Knowledge is the awareness that Frankenstein is *not* the monster, runs a joke; wisdom is the awareness that Frankenstein *is* the monster. Popular culture indeed loves the image of the "mad scientist" whose efforts create unanticipated havoc. Other readings blame Victor's ambition: to create life in the lab without thinking of the rest of the world. These readings, which are behind "Frankenfood" language, treat the monster as symbolic of the breakdown of responsible human stewardship of nature.

In a 2011 article in the *Breakthrough Journal* entitled "Love your monsters: why we must care for our technologies as we do our children," the French philosopher Bruno Latour proposed that the moral of *Franken-*

stein is that technologies should not be conceived, created, and unleashed without human care and guidance.[6] The scientists who discover things like nuclear fission or genetic modification, for instance, are responsible for their applications. This is a parental reading of the novel's moral: constantly care for your creations or you're a bad parent!

Each of these readings has some truth to it, but is implicitly conservative and socially passive. They give us a comforting spectator position from which we can stand to look down on the story and identify what's wrong. But the position of a spectator can anesthetize the onlooker and make the solution seem easy. Centuries from now, no doubt, the textbooks will say how unbelievably stupid and shortsighted twenty-first-century humans were for not addressing the shrinking glaciers and the carbon dioxide they were pouring into the atmosphere. They will shake their heads at parents who did not vaccinate their children, and regard us as fanatics and criminals.

Shelley's story, read attentively, wrecks such interpretations. She leaves no indication that the beast has anything but a normal brain. The creature himself gives a persuasive reason for his behavior: "I was benevolent and good; misery made me a fiend." He was so human, it seems, that he eventually did what anyone would do. Shelley's story indeed carries a shockingly contemporary message: the creature is the extreme refugee, unable to assimilate into the country where he's thrown without his consent, whose inhabitants not only reject but vilify him.

Additionally, in Shelley's novel, Victor's ambitions are not outlandish but shared by the professors who taught him science in the first place. The language Victor and his professors use echoes that of early modern scientists. His desire to "penetrate into the recesses of nature, and show how she works in her hiding places" is Baconian, for instance, while his desire to build his science "upon a secure foundation" is Cartesian. Mad or not, he and his professors are acting like solid citizens of the scientific workshop.

The science, too, was plausible. It is an early fictional example of what Hannah Arendt will call "universal science," in which scientists act "into nature" and science introduces elements of uncertainty and unpredictability into the world. The numerous contemporary fictional examples include ice-nine, the out-of-control material in Kurt Vonnegut's novel *Cat's Cradle* (1963) that was invented by the military to freeze mud so that soldiers could walk on it, but whose potential for freezing water nearly destroys humanity; the film *Blade Runner* (1982), based on a Philip Dick novel about out-of-control, bioengineered humans; and Michael Crichton's thriller *Prey* (2002), about out-of-control nanobots.

A real-life example of acting into nature is the fate of the Mer de Glace, the glacier that Shelley visited just before she conceived *Frankenstein* and on which she staged a scene of the novel. Tourists used to visit the once-mighty glacier for its wildness and majesty; now it's a monument to global warming. The field of ice on which the creature confronted his creator is vanishing, thanks to atmospheric warming following the introduction of large amounts of carbon dioxide into the Earth's atmosphere as a by-product of fossil fuel consumption. What is the cause of *that*? It can't be blamed simply on industrialization, nor on evil, planet-destroying companies. Nor can the blame be placed on the tune-out alternative adopted by many US politicians—to deny or ignore the melting, assuming that the warming will magically reverse itself. The brilliance of Shelley's story is that it thwarts every attempt for the reader to step outside and embrace a spectator position.

Vico had seen the danger of the ability of Cartesian-like science to dazzle human beings into devaluing what binds them into communities, leading to a "barbarism of reflection" in which one can "go mad rationally." Shelley's novel brings to light another danger, the very atmosphere in which science can be conducted. The evil that erupts in Shelley's story cannot be blamed entirely on electricity, careless scientists, psychology, or chance; it also arises in part from the climate in which Victor conducts his research. It recalls another contemporary predicament: that of bank employees whose managers instruct them to handle their accounts ethically but who must work in a banking climate with overwhelming incentives not to do so. Shelley's story rings an alarm bell, but not because of some breakdown in the system—bad behavior, brains, or breeding. What's most alarming is that *nothing* broke down. The wrong research climate, coupled with the vast, powerful, nearly incomprehensible network of human interactions with nature, creates a potential for tragedy as profound as that in any Greek chorus.

Now that's a hideous idea.

SIX

AUGUSTE COMTE'S RELIGION OF HUMANITY

THE TEMPLE OF REASON was locked. I had the right address—7, rue Payenne, a short walk from the Picasso Museum in the Marais district of Paris—but nobody answered the door. A few minutes later Daniel Labreure showed up with a key. The Temple of Reason, he explained, was only open to the public on special occasions.

We walked up a flight of stairs and Labruere flipped on the lights. At first it looked like an ordinary small, private Catholic chapel. The floor was covered by a warm red carpet. The side walls were lined with alcoves topped by Gothic arches, with a bust of a saint atop a pedestal painted in each. An altar at the far end was placed in front of a religious-looking portrait of a woman cradling a child, labeled "Humanity." A flag hung to the right of the altar.

Labreure is the curator of the Temple as well as of a museum, also in Paris, devoted to the French philosopher Auguste Comte (1798–1857), founder of a movement known as positivism. Comte's major works are the *Course on Positive Philosophy* and *System of Positive Philosophy*, and his contributions include numerous words and concepts, such as *altruism*, *positivism*, and *sociology*. In his later years, Comte envisioned a Reli-

Interior of the Temple of Reason, Paris.

gion of Humanity that would replace God with humanity as the object of worship. The Temple of Reason, Labreure explained, had been built by Comte's Brazilian followers according to Comte's strict description. They meant to build it in the home of Clotilde de Vaux, a woman whom Comte idolized and regarded as the inspiration for the Religion of Humanity. But they consulted the wrong document, and built the Temple not in Vaux's apartment at 5, rue Payenne, but next door.

Prompted by Labreure's story, I took a closer look around and noticed uncatholic details. The saints included Archimedes, Aristotle, Gutenberg, and Shakespeare. The painting was of Clotilde de Vaux, painted by a Brazilian artist. The flag was the Brazilian national flag, bearing Comte's slogan "Order and Progress."

Some writers, including Michel Houellebecq, find Comte's message—that the practice of reason needs to be accompanied by some affective involvement, "religious-like," for lack of a better word—inspirational. Others are horrified. The conservative magazine *Human Events* declared that, along with works by Hitler, Marx, and Chairman Mao, Comte's *Course* was one of the "ten most harmful books of the 19th and 20th centuries" because it held that "man alone, through scientific observation, could determine the way things ought to be."[1]

Comte's religious project was a "complete, even preposterous, failure," wrote the intellectual historian Andrew Wernick. What's astute, he continued, was "the thinking behind the project."[2] That thinking provides an important clue for understanding science denial and how to confront it.

BIPOLAR GENIUS

Auguste Comte cut an unlikely figure for a revolutionary, or even the author of a harmful book. Passing him on the street, typically dressed in a neat black frock coat and top hat, you'd take him for a bourgeois conservative. He was short, stout, nearsighted, and clean-shaven. He had a high forehead and dour downturned mouth. Straight eyebrows framed his piercing eyes.

Auguste Comte (1798-1857).

One of the most self-destructive of public intellectuals, Comte also lacked the steady personality that revolutionary leadership ordinarily requires. His biographer, Mary Pickering, called him a "nineteenth-century drama queen,"[3] and another scholar declared him "as pathological an egocentric as ever strutted the stage in a Strindbergian madhouse."[4] Comte actually spent time in a madhouse, and twice attempted suicide. Today he would surely be diagnosed as bipolar; his moods surged back and forth between profound insecurity and the belief he was humanity's savior. He vilified his devoted wife, idolized and abused a younger woman (de Vaux) who struggled to fend him off, and insulted and demanded money of friends and followers. In what he called "cerebral hygiene," he refused to read newspapers and journals, convinced that geniuses such as himself best hatched ideas directly from their skulls. Yet some of his ideas became influential all over Europe and beyond.

Comte was born in Montpellier at the end of the French Revolution and attended the École Polytechnique in Paris, France's leading school of science, math, and engineering. Expelled for helping to lead a protest against an unpopular teacher, he became a political journalist. In 1817, he met Henri de Saint-Simon (1760–1825), a flamboyant and visionary thinker and activist, and a founder of socialism. Approaching sixty, Saint-Simon was the kind of megalomaniac aristocrat—an idealist, eccentric, and over-the-top writer—who enlivens the history of nineteenth-century socialism. He had little scientific training, but confidently spouted grand ideas. He once described a dream in which God said that Newton's law of universal gravitation provided the key to reorganizing human social life, if humans could only find the equivalent law that governed the attraction of humans to each other. To bankroll this and other schemes, Saint-Simon courted powerful and wealthy industrialists and bankers.

For Comte, a scientifically trained, socially conscious nineteen-year-old, meeting Saint-Simon was like encountering an honest-to-God prophet. The pairing was perfect, for a while at least: the aging, vision-spinning poet and the youthful, scientifically minded, energetic social activist. Comte was enchanted by several of Saint-Simon's ideas. One was that the social world is as governed by laws as the natural world, full of "positive content" or demonstrable knowledge.[5] The second was that we seek such laws not for their own sake or for God's, but for humanity. A third idea that Comte surely picked up from Saint-Simon—who in turn was surely indirectly inspired by Vico—was that humanity and human knowledge evolves through three stages (Comte did not discover Vico himself until later), with the final stage a scientific or "positivist" one. Comte made his career goal working these insights into an entire "positive philosophy."

Stone plaque at 34, rue Bonaparte, Paris.

This three-stage idea structured all of Comte's work. Comte claimed it came to him as a sudden flash of intellectual lightning at his apartment one morning in 1822. This is another spurious genesis story. While Comte fleshed out the idea, Saint-Simon helped plant it. Comte's origin story was motivated partly by his self-image as a genius able to summon original ideas from the beyond, and partly by the desire to marginalize Saint-Simon's influence. Today, the legend is commemorated by a stone plaque at 34, rue Bonaparte, where Comte then lived (now the Hotel Saint-Germain-des-Prés).

Over seven years, conflicts grew between Comte and Saint-Simon. Their friendship finally ruptured in 1824, with the publication of Comte's "Plan of the Scientific Work Necessary to Reorganize Society," which included the idea of the three stages of humanity. Saint-Simon wanted to publish it in such a way as to suggest he was the author. Comte insisted on his own byline. Saint-Simon grew petulant, Comte accusatory. Deviously, Saint-Simon published a hundred special copies of the document under Comte's name. But the title page of the real run—a thousand copies to be distributed to influential people—did not mention Comte. The subsequent breakup between the two control freaks was petty and ugly, and it left Comte without a job, income, or colleagues.

Lonely and depressed, Comte married Caroline Massin, an intelligent and witty but penniless woman whom he later said he had first sought out as a prostitute. Grandiose yet insecure, Comte was motivated by the attraction of saving a fallen woman and by his fear that he was too ugly and unsociable to attract any other.[6] For a time, marriage stabilized him. He ate, rose, and retired at regular hours, claimed to be happy, and took on austere habits such as dressing neatly all in black to resemble a priest. Massin adored Comte and looked after him, or tried to, for the rest of his life. But Comte's family loathed her and were outraged by Comte's insistence on a civil rather than a Catholic ceremony.

Comte seemed to revel in poverty and often turned down jobs, such as teaching high school, that he felt beneath him. Boldly, he advertised a lecture series on his "positive philosophy" that the public could pay to attend. He was only twenty-eight, but his reputation as a public intellectual was sufficiently high that subscribers included former students, devoted readers, and well-known politicians and intellectuals. He gave the first lecture on April 2, 1826, at his apartment. Two more followed, but he didn't show up for the next. "He had quite literally gone mad," Pickering wrote.[7] It is unclear what mixture of personal, psychological, and professional problems caused his breakdown, but it was complete. He tried to commit suicide by throwing himself into the Seine, pulling Massin with him. She survived, rescued him, and had him institutionalized. When he was released, still mentally ill, she cared for him at home—where he again tried to commit suicide by slitting his throat. Comte's parents seized the opportunity to carry out what they had always wanted—a formal wedding for their son with the full Catholic ritual. The ceremony satisfied them despite the fact that the groom was raving mad.

Massin slowly nursed him back to health until he could finally return to writing. In 1829 he resumed the lectures. Here's one student's description:

> Daily as the clock struck eight on the horloge of the Luxembourg, while the ringing hammer on the bell was yet audible, the door of my room opened, and there entered a man, short, rather stout, almost what one might call sleek, freshly shaven, without vestige of whisker or moustache. He was invariably dressed in a suit of the most spotless black, as if going to a dinner party; his white neck-cloth was fresh from the laundress's hands, and his hat shining like a racer's coat. He advanced to the arm-chair prepared for him in the center of the writing-table, laid his hat on the left hand corner; his snuff-box was deposited on the same side beside the quire of paper placed in readiness for his use, and dipping the pen twice into the ink-bottle, then bringing it to within an inch of his nose, to make sure it was properly filled, he broke silence: "We have said that the chord AB," &c.

Comte would carry on for three-quarters of an hour until the clock struck nine; then,

> with a little finger of the right hand brushing from his coat and waistcoat the shower of superfluous snuff which had fallen on them, he pocketed his snuff-box, and resuming his hat, he as silently as when he came in made his exit by the door which I rushed to open for him.[8]

Comte turned these lectures into his first book, *Course on Positive Philosophy.* It's an epic: how humanity reached the threshold of maturity, and what it would take to finish the job. From beginning to end, humans find their surroundings at times comfortable and foreseeable, and at times disturbing and threatening; food and shelter appear, but so do floods and hurricanes, disease, and war. What changes during the story is how humans react. The epic took him a dozen years and 4,712 pages to complete.[9]

THE GREAT STORY

"I have discovered a great fundamental law," Comte wrote in the introductory lecture to the *Course*. The law outlines the plot of the story—that humanity, and each branch of knowledge, "passes in succession through three different theoretical states: the theological or fictitious state, the metaphysical or abstract state, and the scientific or positive state." Comte has Bacon's enthusiasm for science, plus Vico's for history. But Comte's interest in history, and the lessons he draws from it, are different from Vico's.

Here is the short version of Comte's epic: in the beginning, humans naturally assume that nature's actions must be due to the wills of powerful spirits. It is gods, for instance, who move planets around the Earth. This is humanity's theological stage. By "theology," Comte does not mean a set of beliefs but a way of life that binds together an explanation for how the world works with practices for engaging it. The theological stage is a complete Theory of Everything, down to nature's smallest details: the spirits are at it again! It also implies a way of coping with nature's threats: placate the spirits! In truth, it's hard to picture nature as run by a gaggle of independent spirits, so animism and polytheism are eventually replaced by a picture of nature as run by a single God. The theological way of life was imperfect and deluded, Comte writes, but an essential step in the development of human thinking. It trained and prepared people—encouraging them to reason and act intelligibly, to try to explain nature's behavior comprehensively, to overcome inconsistent explanations with more coherent ones, and to develop coping strategies—what we now call technology. "For Comte," writes the Comte scholar Robert Scharff, "prayer and ritual form the basis of what is essentially our first technology."[10]

Though the theological stage provides a complete explanation of nature, humans ultimately find that it fails to satisfy. Prayer and ritual don't make nature's disruptions and threats disappear. Repeated failures slowly lead humans into seeing the comings and goings of nature as due not to wills but causes. The planets are no longer seen to be moved by God but by a force called gravitation.[11] This is the metaphysical stage. It, too, involved a Theory of Everything, with "a single great general entity—nature—looked upon as the sole source of all phenomena."[12] The metaphysical stage also provides a complete explanation of nature in terms of a rather mystical ultimate cause. Its theory of everything—"Nature did it!" rather than "God(s) did it!"—gave humanity a better grip on the world and led to better technologies.

But abstract reasoning about nature also has limits, leading humanity into the third, scientific way of life. "The human mind, recognizing the impossibility of obtaining absolute truth, gives up the search after the origin and hidden causes of the universe and a knowledge of the final causes of phenomena." In this stage, humans no longer need to rely on a Theory of Everything, but use the tools and techniques of science. Positive, scientific knowledge about the world and human behavior reduce uncertainties and lessen dangers to human life. Floods, disease, and human conflict still happen, but scientific theories and experiments allow humans to improve their ability to predict them and reduce their impact.

Comte, like Vico, saw the scientific stage not as a break in humanity's evolutionary story, but as the final stage of a continuous process in which humanity transforms its way of life by being attentive to its experiences, learning from its mistakes, and changing the way it goes about things and interacts with the world. Humanity could no more have skipped the earlier stages than mature humans could have skipped childhood and adolescence.[13] You can't fully understand why humanity is in the scientific stage unless you know what it was like to live in a prescientific era, when flashes of lightning sent us diving into caves for protection from the gods.

While Vico told a story of the paths that individual nations take, Comte (like Marx) described a story whose finale unified people all over the globe. In the scientific stage there is only the global human community. Scientists don't have nations or beliefs or principles; they propose theories based on evidence in a publicly shared process that is communal in a way not possible in the first two stages. Humanity in its mature stage is open and interdisciplinary.

Comte ended the introduction to the first lecture of his *Course* triumphantly. "We must complete the vast intellectual operation commenced by Bacon, Descartes, and Galileo," he says. In their wake, physics and astronomy began to reach the positive stage in the seventeenth century, when Newton and others quit asking mystical metaphysical questions about what things like weight and attraction "really" are.[14] Their work was bringing humanity into the stage "that is destined to prevail henceforth, and for an indefinite future, among the human race." Positive philosophy, Comte wrote, constitutes "the only solid basis of the social reorganization that must terminate the crisis in which the most civilized nations have found themselves for so long." He concludes, "The revolutionary crisis which harasses civilized peoples will then be at an end."

This first lecture was an overview. The rest of the *Course* presented a unified system of knowledge consisting of the detailed origins, history, and definitive scientific form of the five traditional sciences: mathematics, astronomy, physics, chemistry, and biology. Comte stumbled here and there. A famous example occurs at the beginning of volume 2, where Comte declares that "we can never know the chemical composition of the stars." That was just three decades before the development of spectroscopy gave astronomers a tool to discover exactly that. Modern readers cringe, too, when Comte praises the "true general spirit" of phrenology, which he finds superior to the "radical vices" of the prevailing metaphysical approach to psychology.[15] Some misjudgments were inevitable, given Comte's ambition and estimation of his own genius.

The *Course* contained two new elements. One was the description of a new science, "social physics" (soon renamed sociology) or the science of human collective behavior. This would complete the roster of sciences and the positive philosophy. Bacon and Galileo had not considered such a science. For Bacon, knowledge of nature is direct and of humans indirect, while Descartes thought human collective behavior was not amenable to methodical treatment. Those earlier thinkers assumed that true experimental science was about nature, not the humans whom they considered a special part of God's Creation. Comte, however, insisted human beings *were* part of nature and therefore as subject to natural laws as anything else.

Social physics would turn the techniques of natural science on the human world. The horrors that France endured during the ten years of the French Revolution starting in 1789—political chaos, summary executions, and other terrors driven by self-interest, ideologies, and fanaticism—as well as the forward-looking Second French Revolution of 1830, showed the need to sweep away superstition and ideology and adopt a social blueprint based on justice and reason. Social physics was born of the idea that it is indeed possible to discover facts about the human world relevant to lawmakers. Learning about nature, Comte realized, will not automatically tell us humans the right things to do. For that, we have to learn as much as we can about ourselves. Just as physics gave us the ability to predict and control nature, so social physics will give us the possibility of predicting and controlling society, supplying the groundwork for true political life. Social physics would provide the data that could fulfill Saint-Simon's dream about God and Newton in which the law of gravitation showed the way to a final social reorganization.[16]

Yet the production of social knowledge was not enough. Another class of thinkers, "positivist philosophers," would have to coordinate and implement the results provided by the sciences. No such class was needed in the first two stages. These two stages had Theories of Everything: God and kings united the theological stage, and ideas like nature and justice guided the metaphysical. In the third stage, this systematic character is missing; there is no obvious guide or "objective synthesis," Comte said, of this knowledge in a way that can be applied for the betterment of humanity. Also, an unavoidable division of labor within the sciences threatens to create a "pernicious influence" of specialization. Positivist philosophers are needed to consolidate and coordinate the efforts of all these special disciplines in what Comte called a "subjective synthesis."

Positivist philosophers were Comte's analogue of Platonic philosopher kings, who would rule knowledgeably and therefore justly. They would meet resistance, Comte foresaw. While the public tends to spontaneously accept the first two stages, humans seem prewired to think that they are special in nature, and resist the truth that they are just another part of it. When a science concerns not planets but people, its subjects come preloaded with theological and metaphysical assumptions, which is why social physics had not emerged before. Additionally, much of the public consists of infants and adolescents who naturally think theologically or metaphysically. Positivist philosophers must play a pastoral role in helping to shepherd humans into this third stage, which they do largely by retelling the Great Story of the three stages, putting people back in touch with humanity's earlier experiences.

CAREER SETBACKS

In 1830, Comte took on this pastoral role himself by offering a free course in astronomy to workers in Paris. He was inspired by the so-called Second French Revolution, in which King Charles X was overthrown and replaced by King Louis Philippe. Though the Revolution replaced one constitutional monarchy with another, it was significant, for it also replaced the idea of hereditary rule by popular sovereignty. Comte saw this as a key step toward social positivism and an opportunity to introduce workers to the general theory of positivism. Thinking that astronomy was the clearest illustration of the three stages, Comte secured a room in the City Hall of the 3rd arrondissement for the lectures. He would continue these lectures for the next eighteen years. Comte also continued to give the lectures that would become the *Course* by subscription.

Meanwhile, Comte kept trying to find a steady job.[17] In 1832, he became a teaching assistant at the École, demeaning work for someone who thought he should be giving the lessons himself. But nothing dented his grandiose ambitions, and he asked French authorities to create for him a special chair in the history of science—which did not yet exist as a discipline—at the Collège de France, without success. In 1836, he got a teaching job at the Institute Laville, a feeder school for the École. The next year, the École hired Comte to be an admissions examiner, which required him to travel around France testing applicants. For a while he loved it. He got to ask tough questions of smart students, show off his knowledge, and be treated with awe.

But as he worked on the final volumes of the *Course*, Comte grew yet more grandiose and insecure. He was humiliated by repeated failures to get a university position, and by the fact that scientists failed to respect him. Trying to strengthen his scientific credentials, he wrote a scientific paper on cosmology to demonstrate the power of his positive approach—

but fell on his face when the paper turned out to be incorrect. He picked fights with friends and with his wife, Massin, whom he treated as a servant rather than an equal, and whom he accused of disrupting his work. He was hurt by the fact that few book reviewers noticed his *Course* volumes. Starting in 1838, he refused to read newspapers and journals, calling his refusal "cerebral hygiene." It was "an effort to protect his ego," Pickering writes, "by turning his back on the contemporary world that he felt was neglecting, if not persecuting, him."[18]

Comte continued to pick petty fights with friends and scholars. In 1835, the Belgian mathematician Adolphe Quetelet (1796–1874), unaware of Comte's work, published a book dealing with the statistics of human populations whose subtitle used the phrase "social physics." Comte accused Quetelet of theft. Out of spite, Comte coined a new name for his special discipline—sociology, the name that stuck—horrifying language purists by conjoining Latin and Greek roots. The contrast between Comte's qualitative approach to human behavior and Quetelet's quite different, strictly quantitative approach is still an important divide in sociology.

But Comte's most self-destructive act was writing the preface to volume 6 of the *Course*. Enraged after getting turned down yet again for a professorship at the École, Comte decided to use the preface to paint himself as a martyr. He ignored Massin's pleas against the idea; she feared it would cost him the lowly École position he depended on. In what even Pickering, the most patient and sensitive of biographers, calls "pathological," Comte accused France's most eminent scientists of conspiring against him and destroying the École. His publisher balked at the defamatory remarks and inserted a disclaimer. Comte sued and won on technical grounds, but was awarded next to nothing in damages. Scientists now openly jeered at him.

Massin, who had just moved with Comte to an apartment at 10, rue Monsieur le Prince, could no longer stand life with her husband, and moved out. But divorce was illegal, and under the patriarchal French law she was still dependent on him not only for money but for permission to move. Comte gave her a small allowance. Still, she attended his lectures, wrote him tearful, loving letters, and often sought unsuccessfully to meet with him.

SOCIAL PHYSICS

The sixth and final volume of Comte's *Course* was published in 1842. The series offered a grand vision of humanity's three-stage evolution. Once sociology is complete, humanity can begin the move into the third, scientific stage. Sociology, again, will not show us how to govern: it will only show us truths about social phenomena—there is no final blueprint for the state. The positivist guardians will rule, using the knowledge that the scientists provide based on the ongoing experiences of the state. But humanity, Comte thought, doesn't enter any stage, even the third, automatically. It has to attend to its experience and learn from its mistakes. Unfortunately, theology and metaphysics persist; humans like the ease and certitude of rules and systems, whether urged by God or Reason or political leaders, and all too often follow authoritarian scoundrels who tell them appealing lies rather than difficult truths.[19]

What made theology and metaphysics so attractive, and what can be done about it? Comte's answer provides one of the great insights into science denial.

In the theological age, Comte thought, God and the King provided the social glue that held society together. Becoming scientific requires letting go of these easily graspable authorities and accepting another, less definite kind of authority, of science as developed by scientists and synthesized by the positivist governors. Moving *from* a theological/metaphysical way of thinking *to* a scientific one is a reflective move, almost a leap of faith to which humanity is driven by the failure of other attempts to advance beyond uncertainty and chaos. Deeply ingrained habits resist such a move. Galileo had encountered such resistance while reporting such straightforward discoveries as that Venus has phases and Jupiter has moons. Resistance to evidence suggesting the advisability of more radical social transformations—evidence, say, that glaciers are melting and the oceans rising—is bound to be far stronger.

It is easy to find such resistance among not only today's evolution deniers but also climate change deniers. "My views on the environment are rooted in my belief in Creation," the popular talk-show host Rush Limbaugh has said. "We couldn't destroy the earth if we wanted to."[20] Representative Tim Walberg (R-MI) once asked himself rhetorically, "Can man change the entire universe?" His answer is another clear example of theological/metaphysical resistance. "No," he said, and continued, "Why do I believe that? As a Christian, I believe that there is a creator, God, who's much bigger than us. And I'm confident that, if there's a real problem, He can take care of it."[21] If glaciers are a principal source of water for His children, Walberg was sure, God would never let the glaciers melt! Consider the remark tweeted by former US vice-presidential candidate Sarah Palin after the 2009 United Nations Climate Change Conference in Copenhagen—"arrogant&naive2say man overpwers nature."[22] That expresses a principle Comte would call metaphysical. Whether "man" can have an impact on nature, Palin thinks, is not a subject for investigation, making it pointless to study glacial melt.

To head off such theological/metaphysical resistance, Comte felt, humans have to periodically reacquaint themselves with the story of how and why they failed to control nature in other ways. This is precisely what Comte saw himself doing in the *Course*.

SCIENTIFIC SEDER

The need to periodically tell the Great Story is one of Comte's great insights. The Great Story is not so much a history of science but of the revelation of the power of science for humans. It is Comte's recipe for keeping dogmatism at bay, focusing humans on their present, inoculating them against the easy answers and tempting principles provided by theological/metaphysical approaches, and reminding them of the historical character of human life.

The *Course* is a kind of scientific Seder. The plot of most Jewish holidays, runs an old joke, can be summarized in a single sentence: "They tried to kill us, we won, let's eat!" Passover gives this an unusual twist. Before participants get to eat, in the Seder ritual, they are recruited to retell the story of the liberation from slavery in Egypt. The retelling, based on a text called the Haggadah ("telling"), is not a history lesson, but a series of reminders and symbols of that liberation. At one point, for instance, the unfolding story halts for a moment of reflection as the participants consider four questions about the meaning of the story. Being part of a religious tradition like Judaism is more than a matter of beliefs or birth within a group whose history includes the Exodus. It means belonging to a community whose members are in danger of forgetting that story and who therefore periodically refresh their memories of it, making that historical experience, and their reflections on it, part of the present moment. Telling the story enriches the tellers' ability to respond to the present. The most radical part of Seder is an imagined conversation between the participants and four children who misunderstand the Seder in different ways: a wise child, a wicked child (a Seder-denier, one might say), an ignorant child, and an apathetic child. This invites participants to grapple with puzzlement, skepticism, ignorance, and indifference. This part of the Seder transcends any strictly religious context, and is important for any group that is exposed to skeptical, ignorant, ideological, and apathetic insiders and outsiders.

The analogy with Seder would not have occurred to Comte, a lapsed Catholic. But both Seder and his Great Story are about communities whose members are constantly in danger of falling out of touch with their own history. Telling the story changes the tellers, refreshing their historical experience, in ways that consulting a history book never could, enriching their ability to think and act.

CAREER SUICIDE

An egotistical oddball in the first half of his career, Comte was brutally weird in the second. Mary Pickering's three-volume *Auguste Comte: An Intellectual Biography* vividly captures his early grandiosity, self-aggrandizement, and misbehavior toward his wife, friends, and publishers in volume 1, which ends in 1842 with completion of the *Course*. In the second volume of Pickering's biography of Comte, Comte's bad behavior and outlandishness are jaw-dropping; it's almost impossible to fathom how someone so nasty, hypocritical, conceited, and self-deluded can be worth reading about. Yet Comte was indeed a visionary who realized just how complex human life is, and how far-ranging preparations have to be to make genuine social change and meet the inevitable resistance.

Insulting his bosses and France's eminent scientists in the preface to volume 6 of the *Course* was just the beginning of his career suicide. In the 1840s, he violated the École's (sensible) policy that forbade admissions examiners from writing their own textbooks, lest students feel pressured to buy them. He wrote a book attacking the École's approach to education—but finally heeding Massin's advice, did not publish it. He ignored his superiors' warnings about complaints of his growing sloppiness and imperiousness as an examiner. In 1843, they considered firing him, but put off the decision. The next year, after he did nothing to change his behavior, and after more complaints, he was fired from his examiner's position (he remained a teaching assistant), to nobody's surprise but his own.

Comte now had no savings, had lost over one-third of his income, and had no employment prospects. For help, he turned to someone who had recently become a close correspondent, the English logician and philosopher John Stuart Mill.

JOHN STUART MILL

Mill (1806–1873) had admired Comte's work, shared an interest in science, and in 1841 sent Comte a fan letter. Thrilled to have what he assumed was a disciple, Comte wrote back a lengthy, self-involved letter about his work and struggles. Patient and polite, Mill sympathized with Comte about what he called the "pedantocracy," a term Comte loved and appropriated. For a while the two got along. Comte quickly grew manipulative, complained bitterly about his poverty, and mentioned his fear that his exposé of the French establishment would cost him his job (which it soon did). Mill good-naturedly assured his new friend that "impartial minds" would support him.[23] Comte began sharing his financial details, including his alimony payments to Massin. Mill took the bait, and offered to help Comte out "as long as I live and have a shilling to share with you."[24] In 1844, when Comte lost his examiner position at the École, he asked Mill to find people to give him 6,000 francs.[25] Mill obliged and got three individuals—two bankers and a former Parliament deputy—to kick in.

The relationship gradually grew rocky after Mill displayed distaste for Comte's enthusiasm for phrenology and for his views on women, whom Comte thought were biologically, psychologically, and socially inferior.

Pickering sensitively traces the evolving, increasingly acrimonious discussion between the two over the "women question," pointing out how deeply personal it was: Comte was tapping his experience with Massin, whom he did not accept as an equal, and Mill was tapping his with his lover Harriet Taylor, whom he did.[26]

Three years later, Comte's views of women began to change when he met Clotilde de Vaux. Then thirty years old, de Vaux (1815–1846) was the wife of a nobleman who had abandoned her and disappeared after stealing money to pay gambling debts. Nearly penniless, de Vaux struggled to become a writer, one of the few respectable occupations open to women at the time. In late 1844, she was introduced to Comte thanks to her brother, who was an acquaintance. Pickering's description of the encounter is priceless:

> He was short, pot-bellied, and balding, with a bothersome strand of hair in the middle of his large forehead. Sometimes when he spoke, a bit of foamy saliva oozed out of one corner of his mouth. He squinted because of his extreme nearsightedness. One eye was sticky and teared constantly. He had various tics, such as twisting his neck from one side to the other. As usual, he was dressed all in black as if to replicate a priest's frock. After seeing him for the first time, de Vaux could not restrain herself from giggling and whispering to her sister-in-law, "He is so ugly! He is so ugly!"[27]

Thus began one of the strangest obsession stories in the history of romance. Comte began tutoring her and writing to her, and made sexual overtures which she repulsed. "The story of Auguste Comte and Clotilde de Vaux," Pickering writes, "is basically the tale of a man trying to force a woman to accept his sexual advances and his desire to be the center of her universe, while she makes every effort to resist him and create her own autonomous life."[28] De Vaux is often dismissed by Comte's followers as a "bland virtuous madonna," and by critics as a gold digger or midlife crisis projection.[29] Pickering shows her to be complex and multidimensional, far more self-aware than Comte. She succeeded in keeping the abusive Comte at bay while managing to learn from him, one of the most arrogant and domineering men of the nineteenth century.

Early in 1846, after knowing Comte for just over a year, de Vaux contracted tuberculosis and began to waste away. On the evening of April 5, Comte barged into her parents' home, pushed into her room, bolted the door behind him to keep out her frantic parents, and waited with her until she died, leaving her family members pounding on the door in fear and anger. In the aftermath, Comte barely avoided a duel with de Vaux's brother.

His encounter—it's hard to call it a relationship—with de Vaux softened but hadn't eliminated Comte's misogynistic views. This grated on Mill. So did an episode involving Comte's paper on cosmology: when Mill heard eminent British scientists dismiss the paper, he assumed Comte was correct and asked him to reply. Comte refused, never admitted his mistake, and accused the British scientists of unprofessional behavior and of conspiring against him.

Meanwhile, Comte's financial situation continued to worsen. He assumed the École would rehire him, and that he could tutor rich students at an exorbitant rate. When neither fantasy materialized, he counted on his three British benefactors to renew their support. They refused; their support had been temporary, conditional on Comte finding other income. Outraged, Comte wrote Mill accusing them of meanness, calling them worse than his French persecutors.[30] Finally angry, Mill reprimanded Comte. Comte delivered another broadside, saying the British were not living up to their "moral obligations" to support him.[31] Now realizing he was dealing with a borderline psychotic, Mill curtailed the correspondence and only sent Comte a few more letters, including a brief condolence after de Vaux's death.

Mill outlined his agreements and disagreements with his former friend in a book, *Auguste Comte and Positivism*.[32] He agreed that scientific knowledge is the only real knowledge, and that humanity is entering a new, scientific age. But he saw no need to tell the Great Story, and disagreed with Comte's view that scientific method cannot be codified into rules. Mill's disagreement was a landmark in the divergence of two views of science. One (Comte's) was that science is a tool for social benefit, though with some potentially negative side effects. The other (Mill's) was that science is an independent intellectual inquiry to be pursued for its own sake.

Why bother to tell the Great Story? Been there, done that, Mill thought. Our main task now, for Mill, was to codify the scientific method, translating it into proofs, rules, and procedures.[33] Mill was what we might call a *nouveau positiviste*, for whom what's important is knowledge, divorced from the cultural and historical milieu in which it originated. Such a person regards "scientific rationality" and "scientific method" as if it were the thinking of a contextless, abstract mind.[34] Mill's view is the key stage in the development (which began with Descartes's image of the view from the vessel far out to sea) of what later would be called the God's eye view of the world—a view from nowhere without any historical sense of how we got there and why we seek what we do in the way we do.

But for Comte, you can't see anything from nowhere. Thinking scientifically requires recalling what it was like to have tried and failed to control nature by the tempting means of prayer, ideology, and reasoning. Mill represented what Comte feared; someone who charged ahead without remembering one's history. When he wrote that Mill was a metaphysician who appreciated "the intellectual value of positivism . . . without understanding its social significance," he was on target.[35] Mill would have been just as baffled by science denial as many of today's scientists—and Comte would have understood why Mill didn't get it.

REVOLUTION OF 1848

In 1846, a depression swept Europe, throwing millions of laborers out of work and sparking antigovernment sentiment. Early in 1848, a series of protests were held in Paris to promote the popular causes. Nervous at their growing scale, the government shut one down, sparking a bloody riot. Louis Philippe, ruler since the Revolution of 1830, was terrified and fled the country in what is known as the Revolution of 1848, the Third French Revolution. A provisional government implemented some long-demanded social measures, such as unemployment aid and the right of all adult males to vote. Despite the progressive schemes, struggles among political factions left citizens afraid that Paris and France would plunge back into the disarray that had been simmering since the 1789 revolution.

Comte was thrilled, calling it "the greatest event to happen in the West since the fall of Bonaparte." Surely this was positivism's—and his—moment! The three French Revolutions were a lesson to France, and the rest of the world, that human society must be reorganized on scientific principles or it will continue to tear itself apart. The revolutionary task begun in 1789 must be completed with scientific principles to propel humanity into a harmonious and progressive future, allowing it to "reorganize without a God or a king."[36] France's leaders, he felt, eventually would have no choice but to turn to him and to positivism for leadership. He alone had the intellectual tools and vision, as well as the backstory for why these were necessary.

Comte established an organization to shepherd in the new era, giving it the hopelessly unwieldy title "Free Association for the Positive Instruction of the People in Western Europe." He soon wisely changed it to "Positivist Club." Such clubs, organized around a political agenda and in effect proto-political parties, had been instrumental in driving the French Revolution of 1789 and were a familiar part of the French political landscape.[37] Karl Marx, also in Paris working on the *Communist Manifesto*, had organized a club called the Communist League. Like Marx, Comte saw his Club as attracting members all over Europe, and as using intellectual and scientific muscle to organize the last social transformation humanity would ever need. Also like Marx, Comte was aware of the importance of popular culture and visual imagery, and promoted his Club through posters, medals, manifestos, and flyers.[38]

While the officially adopted slogan of France in the 1848 revolution was *liberté, egalité, fraternité*, Comte chose "Order and Progress" for his club. Order refers to the understanding of the natural and social worlds; scientific understanding of that order makes progress possible. As he put it in another frequently repeated slogan, "From knowledge comes prevision, from prevision comes action." But while the first two stages sought order based on first principles—theology and ideology—the third stage derives its order from the ongoing pursuit of science and the experiences of life.

HIGH PRIEST OF HUMANITY

One key project of the Positivist Club would be to fashion a "Religion of Humanity" that would replace God with humanity as its focus. This is the strangest part of Comte's thought, but the reasons motivating it demonstrate Comte's insight.

Throughout his career, Comte thought that while science was the path to true knowledge, this knowledge will never come together in a complete package or objective synthesis (as Hegel thought, for instance). Positivists have to create an ongoing and ever-changing "subjective" synthesis of how things tie together in the light of the dangers humanity faces and what is known about the world. In the first phase of his career, Comte felt the Great Story would pretty much serve to get humanity to accept their authority. In the second phase, thanks in part to the impact of de Vaux and the 1848 Revolution, Comte paid more attention to the attachments that lead individuals to care for the social whole. More would be needed, he thought, to motivate people to care for the world, and that care had to be a matter of the heart as well as the head. Their attachments would have to be extensively and systematically cultivated in a set of rituals, practices, and sacraments similar to those used by religions. While in the first phase positivism had simply replaced religion as the social glue, now positivism became in effect a religion itself.

Humanity would have to become the new "Supreme Being" or "Great Being." There will never be a final third stage; there will always be a tendency in any real society for people to be egotistical and greedy, which will have to be countered by a new religion to cultivate love for humanity together with the urge to live differently in a way that will develop humanity. People will have to become emotionally attached to and comfortable with service above their self-interests, for which Comte coined the now-familiar term "altruism."

The Catholic religion, Comte knew from his youth and de Vaux reminded him, reinforced the bond between humans and God with rituals, practices, and other visual culture designed to generate an affective connection in worship and social cohesion among followers. Comte decided that he needed to do the same with positivism. He decided to appropriate and transform for positivism Catholic practices such as sacraments, flags, calendars, holidays, libraries, cults, and so on, all to connect individuals with the social whole, reinforcing its primacy both cognitively and affectively.

In summer 1848, Comte presided over the first of a half dozen or so "positivist marriages" between two disciples. Other projects included the Positivist Library, a core curriculum he drew up for his followers, and the Positivist Calendar, which reorganized the months and years to eliminate religious connections. Year 1789, for instance—the year the Bastille fell—became Year 0. Months were renamed—March became Aristotle—and days were linked with celebrations of important individuals: April 6 was Sainte Clotilde day. Comte also described different kinds of spaces where humans would gather, run by different kinds of people; women would run the salons, for instance. Feminists of the day took notice. Though Comte was both personally and politically a male chauvinist, he at least saw women as having a positive and unique role in his social reorganization. As Pickering writes, "He organized society in an almost maniacal way to create spaces where sociability could blossom."[39]

Eight people attended the Positivist Club's first meeting at Comte's house on March 12, 1848. It met weekly thereafter, and the membership slowly rose to fifty. As usual, Comte undermined his creation through hypersensitivity, paranoia, and delusions of grandeur. That summer he wrote *Discours sur l'ensemble du positivisme*, promoting positivism but dimming its appeal by calling his followers "morally obligated" to support him financially.[40] The Club's membership dwindled. Meanwhile, in the Parisian halls of power, France's leaders never knocked on Comte's door. In a few years, the country returned to monarchy.

Comte's finances sank lower still. Late in 1848, he was laid off from his position at the Institut Laville, leaving him with only 2,000 francs a year as a teaching assistant at the École. Then he lost even that job thanks to a bizarre twist in his ever-bizarre life. For years his École superiors had fielded complaints about him, and Comte somehow survived despite coming within a hair's breadth of getting fired. Then, on March 10, 1851, the École's principal received a routine note from Comte about discount train passes. The principal was about to sign off without thinking—but noticed a strange notation Comte made in place of the date: 13 Aristotle 63. Investigating, he learned that Comte had created a new calendar, and was shocked that Comte used his crazy system on official stationery. The principal was afraid Comte was mentally unbalanced, and fearful of a scandal at the school. Comte was fired shortly after, and was now almost totally reliant on a fund, the Positivist Subsidy, that one of his supporters had created to ensure that his rent was paid.

Comte now set out to write another multivolume opus, the *System of Positive Philosophy* (1851–1854). In that and other writings of the time, he described a "Religion of Humanity" in which he would be the pope. The thought behind it is that science is authoritative because we have experienced it to be the best way to discover the means by which we can predict and control nature, and arrange our society in an equitable way. The rituals and Sermon-on-the-Mount Christianity of Bacon's Bensalem did not come anywhere near what was needed. A new religion was required. What Comte had in mind sounds less wacky if you translate "religion" as whatever you need to do to stimulate humans to do what they are best and most virtuous at, "humanity" as the object of worship, and "worship" as communal expressions of admiration for what is worthy.

Comte's works in this second part of his career are difficult to read. "His earlier writings had resembled the Cartesian order of a well-tended French garden," writes the sociologist Lewis Coser, but "now his work came to look like an untamed tropical forest."[41] He sank further into poverty, and was too poor to afford artificial teeth after his had fallen out. Yet he continued to seek cult status. He commissioned an artist to sculpt his image. He asked friends to address him as "Venerated Priest," and signed his letters "August Comte Founder of the Universal Religion and First Great-Priest of Humanity." He sent advice to world leaders including the tsar of Russia, and the grand vizier of the Ottomans. They never got back to him.

10, RUE MONSIEUR LE PRINCE

Numéro 10, rue Monsieur Le Prince in Paris is a plain, sand-colored, five-story building two blocks from the Odéon metro station. It is in an eclectic neighborhood, next to a Chinese traditional medicine shop and across from the San Francisco bookstore. Comte moved there with Massin in 1841. She moved out the following year, a month after volume 6 of the *Course* appeared with its self-destructive preface. He remained there for the rest of his life, hosting meetings of the Positivist Club, writing various tracts and the *System*, and establishing the Religion of Humanity. The visitors and disciples he met there included several Brazilian students from an École-like university in Rio de Janeiro, who later played a role in the founding of the Brazilian Republic and who purchased both Comte's and de Vaux's apartments (or thought they had), turning the former into a museum and (what they thought was) the latter into a temple.

In 1856, Comte fell ill with cancer. He died in the apartment on September 5, 1857. He contributed to his own decline by rejecting the advice of doctors; they didn't understand the human body as clearly as he could thanks to his positivist principles, he claimed. His will was callous and cruel, all but cutting his ever-loving wife Massin out of support. It contained a sealed "secret addition," to be opened if she contested the will, which accused her of prostitution. Another provision was that his body be left exposed for three days so disciples could see he was really dead.

Auguste Comte's hat and tails, in the Comte Museum.

When Labruere, curator of the Comte Museum, showed me around the place, he said that the apartment has been kept as simple and austere as it was in Comte's time. He pointed out the phrenological skull that Comte consulted in the study, and Comte's top hat and black, priest-like coat still hanging in the closet. A display case contains a locket of his hair, key documents of Comte's positivist movements, and envelopes in which he sent letters to de Vaux. The chair where de Vaux sat when she visited Comte has been deliberately left unrenovated. If you visit, you should be sure to sign the guest book. It will surely be the only time your signature appears in the same document as that of the artist Salvador Dalí.

A COMTEAN NEW ATLANTIS

How seriously can we take Comte's views? He thought that science is the cure for social ills, believed in phrenology but not atoms, did not believe in Neptune or the value of objects outside the solar system, and confidently announced that nobody would ever know the chemical composition of planets and stars. How seriously can we take Comte's claims to being socially progressive given his misogynist behavior? Can we trust someone who thought that a just society can only be held together by a positivist priest-like elite—Comte's own version of an authoritarianism that Marx, meanwhile, was proposing with his "dictatorship of the proletariat"? This runs against modern antielitist sentiments and suspicion of technocracy.

The value of Comte's work, however, has to do with his insights into the complexity of incorporating science into practical human life and of establishing and maintaining its authority. Bacon, Galileo, and Descartes thought that the progressive task was to describe what science was and how it worked; if humanity just got that right, it would mature and know what to do. Comte's experience told him that was not enough. Humanity's maturity is not automatic, and it has to be maintained and reinforced. Comte realized that there's a cost to being interested only in Knowledge and Rationality, for it divorces science from social life in a way that not only reduces the power of science but fosters a collective amnesia that gives rise to social chaos. The workshop then becomes threatened not only by specialization, but also by its lack of connection with the surrounding world.

Comte's New Atlantis would not be a static utopia, but a dynamic community aware that it was in flux from exposure to threats of all sorts. Its citizens would have to care for the community, have that care reinforced, be periodically reminded of what happens when humans do not care, and have experience interacting with those who do not care. The citizens would be periodically reminded of the Great Story to cultivate their appreciation for an experimental approach to meeting threats. They would have habits and rituals to reinforce their attachment to the social whole that went far beyond Bacon's civic Christianity.

But supposing a community had no such religion? Where would the authority of science come from then?

For the writers in Part II, God the Creator was no longer sufficient to establish the authority of those who were investigating it. One might still be able to see the finger of God in the anatomy of a louse, but one could still worry about what the entomologists were up to and the uses to which their work was put. They each told stories to illustrate their fears.

Vico's stories, in *On the Study Methods of Our Time* and the *New Science*, warned how dazzling scientific methods can be and how corrosive these methods were when used elsewhere. Unless these methods were augmented by the humanities, their use tended to make nations grow culturally anemic. Shelley's novel *Frankenstein* foretold that unsupervised scientific work could end up bringing about devastating consequences. Comte realized that science was prone to a debilitating specialization that inhibited its useful application to the world, and that it was not innately authoritative, so that people would have to be cultivated to accept it. But none of these stories provide a completely convincing account of what can be done to fix the danger.

The authors discussed in Part III sought answers.

SEVEN

MAX WEBER: AUTHORITY AND BUREAUCRACY

IN MID-SEPTEMBER 2001, shortly after 9/11, letters containing deadly anthrax spores were mailed to several news agencies and two US senators in a terrorist attack that killed five people and injured more than a dozen others. Unopened mail began piling up in government offices and media headquarters, the terrified recipients fearing for their lives. At the government's request, a panel of scientists researched a method to make mail safe by irradiating it with electron beams. It was a triumph of science's application to national interests.

But when US Postal Service officials implemented the electron-beam method, some batches of mail burnt to a crisp. Its officials had second-guessed the scientists. They had reasoned that if the scientists said the right radiation dose to blitz the death spores was x, then 2x was surely safer! When the dose was scaled back, the method worked.

John H. Marburger, presidential science adviser under George W. Bush who had convened the panel, was alarmed. It was, he wrote later, a "relatively benign example of a potentially disastrous behavior," namely, the tendency of government officials to alter or ignore authoritative scientific advice. Marburger mentioned more damaging examples, including the Bush administration's claim, in 2002, that aluminum tubes sought by Iraq were for a nuclear weapons program, contrary to the conclusion of Department of Energy scientists. In these and other cases, Marburger wrote, "the methods of science [are] weaker than other forces in determining the course of action."[1] In everyday life we often use gut feelings to modify instructions we are given—but shouldn't advice from top scientists have securely grounded authority when requested by the nation's officials? If science is not authoritative, who is safe?

After he stepped down in 2009, Marburger sought the answer by turning to the works of the reigning authority on authority, the German sociologist Max Weber (1864–1920). In his influential, posthumously published book *Economy and Society*, Weber (pronounced *vay-ber*) analyzed different types of authority, or the grounds on which people voluntarily comply with commands issued by others. Suppose humanity never develops something like Comte's Religion of Humanity—what kind of authority can science expect to have? Weber's work provides clues to an answer. It contains key insights into the origins of science denial in the form of a revolt against what he called rationalization.[2] Finally, it eerily anticipated the rudderlessness that would be created by an atmosphere in which scientific authority was low and political authority, amplified by scientific and technological power, was high.

HELLENIC DEPTH

He was actually Max Weber Jr. His father was a patriarchal male who treated his wife Helene like a docile servant, and even refused to let her control her own inheritance. A devout Protestant involved with cultural and intellectual issues, Helene silently endured this treatment. Max Sr. was a member of parliament and intended his eldest child to become a politician like him. For a while in the 1880s, Max obediently followed

Max Weber (1864-1920).

in his father's footsteps and enrolled in the University of Heidelberg to study law and prepare for a life in politics. He joined his father's dueling fraternity, drank more beer than healthy, and gained weight. He did the expected year-long military service, though he found it difficult, and not just because of his new corpulence. He hated basic training and resented how it sought to eliminate independent thinking and to "domesticate thinking beings into machines responding to commands with automatic precision." Yet he also appreciated the results it obtained, turning legions of carefree wastrels like himself into an organized, working machine.[3] This was Weber's first personal encounter with efficiency-seeking, dehumanizing bureaucracies. He would soon research bureaucracies scientifically, and come to see them as driving but also strangling the modern world.

Weber eventually graduated from the University of Berlin and moved back to his parents' home. He became a lawyer and government adviser, and got a PhD in economic and legal history for work that was in effect sociology. Sociology, conceived by Comte, was not yet a fully established field—the first academic sociology department was created at the University of Chicago in 1892. But the French sociologist Émile Durkheim (who established the first French sociology department in 1895) and others had done much to develop quantitative research methods for investigating social behaviors and institutions.

By 1892, when Weber turned twenty-eight, he was a Renaissance man in an overspecialized world—a talented teacher, lawyer, sociological researcher, and government official who threw himself into each role as if it were a full-time job. A workaholic, he still lived at home. A path out opened when he became engaged to Marianne Schnitger, the twenty-two-year-old daughter of a cousin on his father's side. It was an "intense and moral companionship," she later wrote in her biography of her husband. Other biographers seem sure that their marriage was never consummated; Schnitger is silent on that subject. If so, work habits surely played a role. Weber later described himself as needing "to feel crushed under the load of work" and told Marianne that "If I don't work until one o'clock I can't be a professor."[4]

Weber's work habits paid off—for a time at least—as his academic career skyrocketed. In 1894, he became a full professor of economics at Freiburg University. Three years later, he had a choice between two excellent opportunities: becoming a candidate for parliament, or accepting a prestigious professorship in political economy at Heidelberg. He took the professorship, but maintained a foothold in politics as a political consultant. In Heidelberg, he was given an imposing mansion on the edge of the Neckar River, overlooking the city's famous castle and old bridge. He was a towering figure sometimes compared to an Old Testament prophet. "Tall and fully bearded, this scholar resembles one of the German stone masons of the Renaissance period," wrote one journalist. Weber gazed at you from deep inside, and spoke with Hellenic depth and clarity. "The words are simply formed, and, in their quiet simplicity, they remind us of Cyclopic blocks."[5]

That summer, tensions within Weber's household exploded. He invited his mother to spend a few weeks in Heidelberg without his domineering father. His father came anyway. When both appeared at the door, Weber stood up to his father for the first time. The two had a vicious fight, and Max turned his father away. After his mother returned home, Max Sr. refused to speak to her, and a few weeks later died of a hemorrhage. Blaming his father's death partly on himself, Weber fell into depression and an illness that left him unable to resume a full academic life for twenty years. He tried to resign from his professorship, but Heidelberg officials were reluctant to let go of such an important man and simply gave him a leave. For the workaholic Weber, taking a university salary without doing much to deserve it was mental torture and depressed him further. He spent time in a sanitarium, staring blankly out the window for hours at a time. Marianne did all she could to help and frequently took him on trips to distract him.

In 1903, things began to look up. A colleague preparing to take over a journal, *Archiv für Socialwissenschaft und Socialpolitik* ("Archives for Social Science and Social Welfare"), asked Weber to be a coeditor. This was the perfect therapy. Never one to go halfway, Weber threw himself into the project, helping to turn it into the most prominent sociology journal in Germany until the Nazi takeover in 1933, when the journal was forced to close. Weber wrote an article for every issue of the initial volumes, plus an introduction explaining the journal's direction. A basic assumption of that direction was that genuine scientific research was value-neutral. This would prove a key concern for science deniers.

GALILEI OF THE GEISTESWISSENSCHAFTEN (HUMAN SCIENCES)

In principle, the value neutrality of science is easy to explain. Think of a polling place. These institutions are set up to promote democratic values and are agents of such values, while those who work in them are expected to practice values like impartiality and civic-mindedness. And when it comes to actually tallying ballots (or, today, operating the vote-counting machine), they are expected to be utterly neutral and objective.

Scientific workshops, in Weber's eyes, operate in much the same way. Much of everyday life is concerned with seeking valuable things such as comfort, health, peace, stability, and wealth. Our aims in these searches are more often than not unrealized, challenged, or threatened. Death and disease menace our desire for comfort and health; poverty, inequality, and injustice obstruct our pursuit of social values. The sciences—both natural and social—spring from our desire to find better ways to realize these values. Medical science, for instance, seeks to develop cures for diseases, while social science explores methods for making good policies.

Science is therefore what Weber calls *value-relevant*. It not only springs from values, but also is itself a value in that it creates a special set of realms—laboratories or workshops—in which scientists can study, clarify, and improve a particular kind of activity. Science is not the only value. One person may respond to an epidemic or to poverty by scientifically studying causes and treatments in a laboratory or scientific workshop, while others may follow more practical paths outside the lab by becoming doctors, social workers, or crusading journalists. But the scientist's work *inside* the lab or workshop is different. It is most effective if it is *value-neutral*, that is, if it abstains from judging how the research will be used *outside* it.

Weber's assertion of the value neutrality of science was controversial at a time when scientists, both social and natural, felt compelled, for patriotic reasons, to use their work to promote and defend their kaiser and the German Fatherland. As the noted German-American sociologist Lewis Coser, founder of the Brandeis University Sociology Department, once wrote:

> It is against this prostitution of the scientific calling that Max Weber directed his main effort. His appeal for value neu-

> trality was intended as a thoroughly liberating endeavor to free the social sciences from the stultifying embrace of the powers-that-be and to assert the right, indeed the duty, of the investigator to pursue the solution to his problem regardless of whether his results serve or hinder the affairs of the national state.[6]

Is such objectivity really possible in sociology? The natural world—the world of atoms and molecules, sticks and stones, and heavenly bodies—stays the same; we humans only study it differently, with ever-improved ideas and instruments. But the human world, as Vico pointed out, is not like that. Humans are not like particles and planets; they produce and act on their own meanings in an ever-changing, self-created world. This makes human history, Weber wrote in a dramatic metaphor, like a murky and swiftly moving river without logic or direction, a "stream of immeasurable events [which] flows unendingly towards eternity."

Studying events in that murky river requires a special conceptual tool that Weber called an "ideal type." An ideal type is a pure construct of a phenomenon that sociologists find useful in studying actual, concrete cases. "Ideal" here does not mean "good." There are ideal types of prostitutes, hackers, and thieves; and of banks, monarchies, and religions. Ideal types are simply abstract measuring sticks that can be used to describe and compare human creations.[7] They make possible what Weber called an "*empirical science* of concrete *reality*,"[8] and their creation proved so useful in the human sciences that Weber has been called the "Galilei of the *Geisteswissenschaften*," the Galileo of the human sciences.[9]

Equipped with such measuring sticks—ideal types of social action—Weber began to explore what he saw as the increasing rationalization of modern life, accompanied by its growing bureaucratization and particular kind of authority. A key step, both in his recovery and in his research, was a trip to America.

ST. LOUIS

In 1903, Weber was invited to the Congress of Arts and Sciences in St. Louis. This was part of the famous St. Louis Exposition of 1904 to celebrate the centennial of the Louisiana Purchase. Weber had never been to America, and would have an opportunity to conduct research and drum up articles for the journal. He and Marianne left for America by boat in August.

The United States was still an exotic place to many Germans, who tended to think of it as uncouth, with dirty cities swarming with a range of ethnicities, mostly immigrants from the poorer sections of Europe. But Weber was intrigued. In New York, their first stop, Weber crossed the Brooklyn Bridge, from which he surveyed the "fortresses of capital" on Wall Street, modern-day versions of the old merchant buildings of Bologna and Florence. From the Bridge's footbridge at rush hour he heard "a constant roaring and hissing; the rattling of the trains is punctuated by the tooting of the steam whistles of the big ferries far down below."[10]

After New York, the couple went to Buffalo and then Chicago. Chicago's semifeudal political scene, with its extensive patronage system mixed with a ruthless quest for money, Weber thought, made Chicago embody even more of the American spirit than New York. He paid a boy fifty cents to give him a tour of the stockyards, trying not to get buried in filth as he waded through the "atmosphere of steam, muck, blood and hides" while following "a pig from the sty to the sausage and the can." Weber noted the cold calculus of a streetcar company that continued to dole out money—$5,000 or $10,000, depending on the accident—to each of four hundred accident victims or their heirs per year, because continuing to pay out that way was cheaper than fixing what caused the accidents. American life, as Marianne put it, had a "magnificent wildness" full of extravagances and wastes, but was also a "monster which indifferently swallowed up everything individual."[11]

After Chicago, the Webers went to St. Louis for the Congress. The

Exposition of 1904 was the largest world's fair to date. Over seven months, 20 million people attended exhibits from sixty-three countries. John Philip Sousa's band played at the fair, Scott Joplin wrote a piece for it, and it inspired the movie and song *Meet Me in St. Louis*. The Exposition's quieter, scholarly side was its Congress of Arts and Sciences, designed to highlight cultural and intellectual progress in the past century. Weber, now forty, was among the lesser known of the foreign scholars in attendance, in the United States at least. The work for which he would shortly become famous, *The Protestant Ethic and the Spirit of Capitalism*, was just then being published in the first issues of the *Archiv*. On the afternoon of September 21, 1904, Marianne was thrilled to see her husband calmly and confidently deliver the first talk he had given in over six years.

The couple headed west from St. Louis. At one point, they came across an oil town that had sprung up in the middle of the Oklahoma forest. Weber noticed that the town's free and easy wildness, its informal officials and irreverent storytellers and jokesters, stood in sharp contrast with the bureaucratization that he saw engulfing other parts of the country. He lamented that, in a year, the place would probably look like any other American city. "With almost lightning speed everything that stands in the way of capitalist culture is being crushed."[12] Throughout America, in fact, he saw powerful tensions between top-down versus bottom-up processes. He called it the "Europeanization of America," or the tendency of top-down bureaucracy to prevail.

PROTESTANT ETHIC

The year after Weber's American trip, 1905, the *Archiv* published the second and final installment of his work *The Protestant Ethic and the Spirit of Capitalism*, one of the most influential sociological treatises ever written. Its influence is due less to its thesis, which has been challenged and modified, than for the way he demonstrates the usefulness of ideal types in sociological research.

The Protestant Reformation, Weber thought, was not a historically predetermined outcome of economic changes, but a matter of humans thinking through how best to live, in which their values played a driving role. Living virtuously, for Protestants, did not mean simply amassing wealth or living in monastic solitude, but involved a "calling" in which they had to engage in "the market place of life" to acquire and maintain wealth in a disciplined and ethical way. This "Protestant ethic" was not individualistic wealth-seeking, but a conscientious, interactive approach to creation in which practitioners engaged in an organized accumulation and reinvestment of capital in a system that relied on free labor rather than peasants or serfs.

Weber also saw capitalism, given moral energy by the Protestant ethic, as dovetailing with another process which he called rationalization.

Like Vico and Comte, Weber saw rationality as a basic human impulse ever since primitive times. Crudely, rationality refers to what drives human society forward, the process by which humans systematically seek to understand and control the world. Primitive humans were rational in a practical sense when they used prayer and ritual to try to get favors from the gods—an early form of investment strategies and portfolio management. But something began happening in the scientific revolution that gave rationality new power and ruthlessness, transforming and interlinking different areas of culture and social life and causing them to operate with a new inflexibility. Weber called this new, virulent strain of rationality *rationalization*. In law, it drove the replacement of wise, Solomon-like judges by formal legal codes. In education, it drove the replacement of apprenticeship and mentoring with state-approved curricula. In telling time, it drove the replacement of sundials and church bells by standardized time dictated by coordinated clocks.[13] Weber saw rationalization as a corrosive but dynamic and open-ended process that was injecting "rational calculation" and the rule of experts throughout culture and social life, squeezing out traditional values and practices.

Weber found rationality present throughout even religious experience, involved in the transition from polytheism to monotheism and from saintly leaders to church bureaucracy, with the Protestant ethic representing the full rationalization of the religious life. But he saw rationalization ultimately as corroding the Protestant approach to capitalism, slowly absorbing and evaporating the religion's moral energy. Early forms of capitalism were open to pricing goods based on convention and other traditional motives, but the sole end promoted by rationalization was maximizing profits. Thanks to it, "material goods have gained in increasing and finally an inexorable power over the lives of men as at no previous period in history."[14] In the United States, he continued, the pursuit of wealth was becoming stripped of religious meaning and was now associated with "purely mundane passions," and even "the character of sport," as when people brag about triumphing over opponents in the marketplace. "The idea of duty in one's calling prowls about in our lives like the ghost of dead religious beliefs," Weber wrote.

Is there a way to restore values and morals to the modern capitalist machinery? Weber was dubious.

> This order is now bound to the technical and economic conditions of machine production which today determine the lives of all the individuals who are born into this mechanism, not only those directly concerned with economic acquisition, with irresistible force. Perhaps it will so determine them until the last ton of fossilized coal is burnt.

Here Weber writes with his inexorable and compelling "Cyclopic" language, and his words abruptly turn ominous. The calling has become a jail sentence. According to one theologian whom Weber cites, in medieval times the care for external goods lies on shoulders of a saint "like a light cloak, which can be thrown aside at any moment." Not for the modern person. Care for such goods has entrapped us, and "fate has decreed that the cloak should become an iron cage." Can we free ourselves from the cage?

> No one knows who will live in this cage in the future, or whether at the end of this tremendous development entirely new prophets will arise, or there will be a great rebirth of old ideas and ideals, or, if neither, mechanized petrification, embellished with a sort of convulsive self-importance.

Then, with a sentence that makes a reader shiver:

> For of the last stage of this cultural development, it might well be truly said: "Specialists without spirit, sensualists without heart; this nullity imagines that it has attained a level of civilization never before achieved."

This was Weber's melancholy prognostication of the looming expert-dominated age.

BUREAUCRACY

After *The Protestant Ethic*, Weber slowly reengaged with academic life, though without a full teaching load. His house became a salon frequented by leading philosophers, sociologists, and historians. Early in 1909, he had a relapse that caused him to stop working. But a few months later, a meeting of the Social Policy Association in Vienna gave him a forum for expressing something he cared about—his ambivalence about the growing impact of bureaucracy. Marianne noted his surge of interest in preparing for the gathering: "He was like a dammed-up stream of intellect that cannot stop flowing and carrying people away."

Weber's energy was fueled by anger at colleagues who celebrated bureaucracy, calling it indispensable for collective action, state socialism, and the improvement of Germany. "No machinery in the world works as precisely as this human machine, the bureaucracy," Weber conceded in an hour-long speech. "From a technical and material point of view it is unsurpassable." Yet it had downsides. For instance, it created dependent and servile individuals. "Everyone who integrates himself becomes a little cog in the machine, just as in a big industrial enterprise, and he is increasingly attuned to feeling like one and to asking himself whether he cannot become a bigger cog." Weber emphasized that he was not objecting to the value and efficiency of bureaucracy, only protesting its uncritical glorification. "The question is what we have as a *counterpoise* to this machinery so as to keep a remnant of humanity free from this parceling out of the soul, from this exclusive rule of bureaucratic ideals of life."[15]

In 1909, the publisher of the *Archiv* asked Weber to edit an encyclopedia on political economy. The result, as Marianne later put it, "unintendedly grew into the major work of his life." As editors must, Weber had to twist the arms of friends and associates to line up volumes, and gave each author a two-year deadline. Much of his own contribution had to do with how economy and society were being reshaped by rationalization, and it slowly grew in size and scope until he, too, missed his own two-year deadline. His effort eventually became a separate book, *Economy and Society*. While Weber complained that he should never have taken on the project, Marianne was thrilled that he was under the spell of a great and challenging task.

In 1914, Germany entered World War I. Germans assumed their victory would be swift and easy, and finally bring Germany the international recognition it deserved. Initially enthusiastic about the war's prospects for reinvigorating Germany, Weber signed up for duty. But he was fifty, and considered no longer fit to bear arms, so he was put in charge of establishing and managing a set of hospitals around Heidelberg.

Once again he was part of a military bureaucracy, though now as a manager. Again he was ambivalent about it. He understood the indispensability of bureaucracy for administering goods and services on a large scale. He admired its goal of fairness and accountability, and its clear jurisdictions and organization of expertise. But he was also frustrated by the inevitable inertia and inefficiencies. Growing suspicious of German ambitions to annex foreign territory, and seeing how unprepared Germany was for a long war, Weber's enthusiasm evaporated, and he returned to research.

Weber's scientific interests had always been motivated by political issues; his interests were always, in his terms, value-relevant. The war now provoked deep concerns in him about the future of Germany, after its strong-willed leader Bismarck had been replaced by the weaker and more impetuous Wilhelm II. These concerns led Weber to consider in *Economy and Society* the elements of a good and stable state, as well as ideal types of bureaucracy and leadership. Bureaucracy is inevitable, he wrote:

> The more complicated and specialized modern culture becomes, the more its external supporting apparatus demands the personally detached and strictly "objective" *expert*, in lieu of the lord of older social structures, who was moved by personal sympathy and favor, by grace and gratitude. Bureaucracy offers the attitudes demanded by the external apparatus of modern culture in the most favorable combination.[16]

But bureaucracy also makes rationalization incarnate, becoming its worldly arms and legs. "The more the bureaucracy is 'dehumanized,' the more completely it succeeds in eliminating from official business love, hatred, and all purely personal, irrational, and emotional elements which escape calculation."[17] Bureaucracy also replaces old-style rulers who are motivated by concern, sympathy, grace, and personal judgment, with emotionally detached, professional experts. Rationalization also corrodes the value neutrality of the scientific workshop, by making it easier for scientists to leave their humanity outside the lab, acting not as scientists but as agents of funding agencies or political ideologies.

Some of bureaucracy's acts of dehumanization can be blunted by what Weber called "ideological halos" that religion, politics, or some other activity occasionally give to certain cultural values. Yet some stronger countermeasure is needed to keep bureaucracy from stifling the modern state. Weber could not find it.

SCIENCE AS A VOCATION

In 1917, as the war neared its end, Weber gave a lecture in Munich on "Science as a Vocation" as part of a series of talks by various professors on different careers that were popular choices of students. He began by characterizing the institutional context—the "game" careerists had to play—followed by an assessment of how the game was changing those who played it.

Weber was discouraging. It was an ominous time in Germany. Three years into the Great War, Germany and the Central Powers were scoring key victories, but the United States had just joined the fight on the Allied side, and enthusiasm for the war was fading due to mounting casualties and scarcer resources. Other professors might have roused students or at least motivated them by recalling past glories of German science and politics. Not Weber. Speaking from just a few note cards, he struggled to adopt a neutral and objective posture, but couldn't keep the gloom from his language as he fulfilled the responsibility of a teacher to, as he said, make students recognize certain "inconvenient facts" that deviate from the party line.

He was older now, and his physical appearance had changed dramatically since the days when he had been compared to a rugged stonemason or artist; now he resembled more the weary soldier. "His face, with a shaggy beard growing all around it, recalled the mournful glow of the Bamburg prophets," wrote one listener, while another described him as looking "like a medieval warrior before leaving for battle."[18] Furthermore, his Cyclopic blocks of language were now more like hand grenades; one audience member described his talk as a "confession" that "burst from the speaker's breast in jerky explosions."[19]

Few students hearing "Science as a Vocation" could have been thrilled when Weber described a German academic career as a crapshoot. With no job security and little pay, it's only possible for trust-funded kids. In the United States, the system is more regular and fair, but rewards second-raters. If you can't stand seeing mediocrities advance year after year, he said, the academic life is not for you.

What could draw a person to science as a calling? Here Weber was even gloomier. Once upon a time, science was a path to God, the quest for traces of His plans and of one's place in the world. For Bacon, Galileo, Descartes, and others, science was a means of taming, controlling, and exploring a God-created cosmos. Science, the revelation of mysteries, was therefore a personal experience that could produce "passionate devotion" and "strange intoxication." No longer. Like bureaucracy, it now served the process of rationalization that had been absorbing Western culture. The gods, demons, and magical spirits were gone; only particles and forces were left. Science was now a career for specialists whose work would be made obsolete by successors.

The specialization of science has several consequences. To be productive, today's scientists have to "put on blinders" and focus on narrow questions. Scientists, for instance, are no longer in touch with the infrastructure, technical and administrative, in which they work, often knowing as little about the big instruments they work on—think of today's particle accelerators and synchrotron light sources—as ordinary streetcar riders know about the operation of that vehicle. In principle, specialists know they *could* understand this infrastructure, the way that riders could understand how a streetcar works by taking intro level physics classes. "The savage knows incomparably more about his tools," and why and how their use relates to values and the meaning of the world.

That was the gloomiest part of Weber's message: "there are no mysterious incalculable forces" anymore in the world, for one can "master all things by calculation." Everything you need to know can be found in those science courses that bored you in high school. The world, Weber said in one of his bleakest and bluntest remarks, has become "disenchanted."[20]

> The fate of our times is characterized by rationalization and intellectualization and, above all, by the "disenchantment of the world." Precisely the ultimate and most sublime values have retreated from public life either into the transcendental realm of mystic life or into the brotherliness of direct and personal human relations.[21]

As for modern capitalism, so for modern science: passion and spirit have been squeezed out. The theories of modern science seem "an unreal realm of artificial abstractions" rather than a way to God or even Nature.

So what is its value? Here, too, Weber had little to offer the students. Science has a certain practical value in its ability to give us things we need, though he said he personally found a career doing this about as exciting as being a grocer selling cabbage. Another possible attraction is that science provides new tools and ways of thinking—still, this is like improving the grocer's methods. A third attraction is that science, while not establishing values, very occasionally can help us achieve clarity about our means and ends. Medical science, for instance, seeks techniques to better cure diseases but provides no guidance for "whether life is worthwhile living and when." Science, in short, is no longer a grand calling. It no longer provides its own moral justification, and is a matter for specialists and experts immersed in vast bureaucracies. To face the key moral problems of the day, you have to turn elsewhere.

POLITICS AS A VOCATION

Weber was feverishly active in the last years of his life, and finally returned to his workaholic and overcommitted habits. He was even politically active, and his name was floated as a possible cabinet member in the new government. But in 1919, falling ill, he withdrew from politics, wanting to focus on the still-unfinished *Economy and Society* volume.

That year, Weber gave another talk in the career series, "Politics as a Vocation," which has been described as his "swan song." In it, he draws on his deep experience with politics and science. He begins by describing the function of institutions in politics: they allow human beings to dominate other human beings for a purpose. Such domination can involve three different kinds of authority, or grounds by which people voluntarily comply with commands issued by others. These three ideal types are traditional, rational-legal, and charismatic.

Traditional authority is rooted in a belief in the sanctity of age-old rules and the "way things are done"; it's the authority of parents and village elders. Rational-legal authority is the authority of a bureaucracy; it's grounded in a belief in the legality of enacted rules and represents obedience to a system rather than an individual. Charismatic authority is exerted by people who are considered to be extraordinary and to possess special powers; think of Martin Luther King, Mahatma Gandhi, or Winston Churchill. Charismatic authority is hard to sustain and requires occasional proofs of special powers, such as an ability to do amazing things or to disclose secrets of nature. It inevitably ends up being transformed into bureaucratic authority; Jesus's message is taken over by the Church, revolutionary leadership by political parties.

Weber called charismatic authority irrational, but said it's one of the few means that leaders have to take groups of people on new paths. As he put it brutally and a bit oversimply to a German general and nationalist leader, "In a democracy people choose a leader in whom they trust. Then the chosen leader says, 'Now shut up and obey me.' People and party are no longer free to interfere in his business."[22] But this irrationality has a positive side, for charisma provides the only counterforce to rationalization. Charisma, one might say, is the momentary irruption of enchantment back into the world, creating a new value—following the leader. The modern world, gripped as it is by rationalization, has outsourced its thinking to the methods, rules, and logic of the technocratic paradigm, driving out personal experience. Weber saw only one way to seize the world back—through the charisma possessed by a leader who happens to be farsighted. This was dangerous, for if the leader is not, things only get worse.

These three kinds of authority are ideal types. No leadership situation involves a pure example of any one of them, but exhibits aspects of all three. Still, analyzing authority in this way allows us to identify key features and conflicts of leadership. An effective leader must curry favor among people charismatically by promising to take them toward social goals, work within the "machine"—the legal-rational authority of the bureaucratic state—and rely as much as possible on "the way things are done."

Weber's analysis also exposes a conflict between an "ethic of responsibility" and an "ethic of conviction." An ethic of responsibility is motivated by values but pays careful attention to the consequences of actions taken to realize them. An ethic of conviction, by contrast—the attitude of single-issue voters—seeks to realize a value come hell or high water, regardless of the consequences, which allows turning the quest for its realization into a matter of calculation.

Weber therefore saw ordinary political action, motivated as it is by an ethic of responsibility, in an "antagonistic interdependence" with the instrumental rationality of science. No single key can resolve the tension between technocrats and leaders: the experts in the workshop, passionate about their work but seeking to be objective and value-free in carrying it out, versus leaders who have the desire to realize worldly goals. Each has potential grounds for being suspicious of the other—politicians seeing scientists as uncaring and detached from the values of the world, scientists seeing politicians as ideologically committed, as not respecting the values of the workshop, as "denying" science.

Early in 1919, at the age of fifty-five, Weber got a chair of economics and sociology at Munich. It was the first chair he had had since his breakdown almost twenty years before. He scurried to complete *Economy and Society*, which contained fuller discussions of his views on authority and bureaucracy, revised *The Protestant Ethic*, and for a while tried to involve himself in political activity. Two semesters later, he caught pneumonia and, on June 14, 1920, he died. He left *Economy and Society* unfinished; the portions published later were put together by Marianne, who also continued the Sunday salons that were now famous throughout Germany.

MARBURGER AND WEBER

When the US Presidential Science Adviser John H. Marburger turned to Weber's writings on authority to understand why government officials had felt free to alter recommendations made by top US scientists, he approached the subject as a theoretical physics problem. Which of the three types of authority, Marburger asked himself, best applied to science?

Clearly the authority of science was neither traditional nor legal-rational, Marburger thought; no society, after all, is traditionally scientific or requires that its laws be grounded in sound science. He concluded that the authority of science in government circles must be charismatic. Science's authority among politicians, that is, depends on their regarding it as possessing a special power or magic. Give scientists enough resources and they can get us to the Moon, cure disease, and create better weapons. More charismatic glaciologists, it seems, are the key to marshaling the political will to stop the melting of the Mer de Glace.

Marburger noted that scientists were bound to find such a conclusion absurd. For them, science is not charismatic but rather the only means at our disposal for reaching an understanding of nature. Because scientists have firsthand experience of how science works—of its roots in empirical testing, open discussion, and peer review—they see acting against science as "a mild form of insanity." Marburger wrote, "It is precisely because the power of science does not require charismatic authorities that we should trust it to guide our actions."[23]

Still, Marburger's conclusion that the authority of science in politics depends on charisma is not far-fetched. From the standpoint of politicians lacking such firsthand experience, the voice of a scientist is but one among many clamoring to be heard. For politicians, "science is a social phenomenon with no intrinsic authoritative force," which is why "the authority of science is inferior to statutory authority in a society that operates under the rule of law." Marburger's observation also explains much of the waxing and waning of science's authority in politics. When scientists make dramatic, socially communicated breakthroughs, their authority shoots up; in years when they don't, it tends to decline. That is why, for instance, physicists had such political power after the Second World War. Identifying science's authority as charismatic also suggests that the only way to garner more authority for scientists in government is to improve the charisma of their calling. For this reason, Marburger concluded, "science must continually justify itself, explain itself, and proselytize through its charismatic practitioners to gain influence on social events."

But Marburger misunderstood Weber. Weber's tripartite scheme concerns ideal types, not actual cases of authority, which blend the three. These types are not categories into which actual situations are to be binned, but tools that need to be used discerningly to understand such situations. Scientific authority has to be embedded, at least in part, in traditional practices and legal-rational structures. Even among scientists, the way things have been done in the past is authoritative and involves some legitimate rule-following.

Weber was aware, too, that there are sources of tension between the three types of authority. When charismatic actresses like Jenny McCarthy champion the antivaccination cause, they get pushback on both traditional and legal-rational grounds of authority. Attempts to change the legal-rational status of mammograms or prostate exams likewise have encountered pushback from traditional authorities. Weber is also talking about authority in a general context, not just in controversies in government circles. The authority of science permeates all decisions with a scientific-technological dimension: a doctor advising what medicine to take, a consultant telling a planning board what bridges and hospitals to construct.

WEBER ON THE MER DE GLACE

Early pioneers of science such as Bacon, Galileo, and Descartes assumed that when people finally understand what science is, its findings will be instantly authoritative: if we know enough about God's Creation, it will tell us the right thing to do. None of them would have conceived it possible, if humanity were to discover that its actions were threatening its own welfare, that this discovery would not motivate humanity to do something about it.

Weber's work explains much about why it might not. First, it makes us realize that, while the controversy over global warming may look like a conflict between individuals—between specific scientists who claim that it is happening, and specific politicians who deny it—this is not so. What's really happening is a collision between bureaucracies that are responding to different constituencies. Second, Weber's work makes us realize that authority is a complicated blend of factors. Glaciologists cannot expect to prevail on charisma alone. In Marburger's anthrax scenario, had the scientists not only provided Homeland Security with data on the antianthrax method but enlisted superiors in the postal service to stage demonstrations of the method at traditional meetings, this convergence of different types of authority would have greatly strengthened the authority of the recommendation.

Weber also reminds us that expert advice is authoritative only when perceived as relevant. When a sick person visits a doctor, an architect consults an engineer, or a policymaker a sociologist, the expert delivers advice to someone prepared to find it relevant; "hot" or active expertise, we might call it. But someone far from the Mer de Glace may not care about its melting; for them information about what is needed to stop it is "cold" or passive expertise. Scientists may feel that, for those who reject science, the ethics of responsibility has been replaced by an ethics of conviction. But "scientific truth is only valid for those who *seek* truth," as Weber wrote in "Science as a Vocation." You have to care about the numbers first before seeking them. Otherwise, those who tell you what the numbers are may appear to be operating politically, out of their own interests.

Most insightfully, though, Weber's work allows us to understand the vehemence with which science deniers often respond to the scientific establishment. They are objecting to the iron cage—to what they perceive as an empty instrumentalism, to the wholesale technologization of life, to an impending era of "specialists without spirit, sensualists without heart"—and want to resist it. Science deniers may know that it is not literally true that global warming is a hoax, but they may feel that efforts to combat it are squeezing out other essential values like jobs and freedom of action. Antivaxxers may know that vaccines do not cause autism, but they may feel that vaccinations are one important way that the medical establishment extends its grip over their children, tearing them away from their parents. Such people may be willing to say something like, "I am not a scientist" and sign on with a charismatic leader who denounces the iron cage, finds science as the obstacle and as politicized, and sees no alternative to denouncing it to protect true human values. Nowhere was this tension between science on the one hand, and cultural values and practices on the other—and whether the former necessarily undermines the latter—more directly and explicitly faced and debated than during the late Ottoman Empire, during Weber's own lifetime.

EIGHT

KEMAL ATATÜRK: SCIENCE AND PATRIOTISM

*For everything in the world, for civilization, for life, for success, the truest guide is knowledge and science (*ilm *and* fen*). Searching for other guides is unawareness, ignorance, misguidedness.*

—KEMAL ATATÜRK (1924)

MUSTAFA KEMAL ATATÜRK (1881–1938) was the founder and initial president of the Turkish Republic, the first secular republic in a Muslim-majority nation. In 1924, the year he abolished Islamic rule in Turkey, his catchphrase about the necessity of knowledge and science as a guide was controversial and courageous. Though the catchphrase is engraved on the walls of many schools and administration buildings in Turkey today, it remains controversial among certain segments of that society.

Atatürk's catchphrase was only possible thanks to a century-long cultural transformation. Before it, religion had permeated all aspects of life in that region of the world, with little or no Western-style boundary between the secular and the spiritual. Changing that atmosphere so that science could be authoritative involved developing new notions of God, country, citizenship, and patriotism. That process was much different from the church-versus-state struggles in the West, yet reveals much about Western-style science denial and how to confront it in the West.

ILM AND *FEN*

Until the beginning of the twentieth century, the Mediterranean inner basin region, which included much of southeastern Europe, western Asia, and northern Africa, was ruled by the Ottoman (or Turkish) Empire, a multiethnic state whose capital was Constantinople (Istanbul) and whose official religion was Islam. During the reign of Suleiman the Magnificent (1494–1566) and for two centuries after, the Ottoman Empire was a powerful economic and political rival of European powers and the Russian Empire.

But a series of costly military defeats in the mid-eighteenth century, most dramatically in the Russo-Turkish War of 1768–1774, made it clear to Ottoman rulers that their empire was in jeopardy. The danger had nothing to do with God punishing them for lack of religious zeal; their enemies, after all, were heathens. Their enemies were evidently capitalizing on some "new knowledge" involving special military know-how. In the 1700s, Ottoman administrators had made sporadic attempts to import the Europeans' special knowledge, but that was difficult in a country where traditions ran deep and strong. A School of Military Engineering, for instance, was created in 1734 to promote European-style artillery practices, but was soon shut after running afoul of religious traditions.

Acquiring the "new knowledge" could not be done piecemeal. You could not select certain bits and pieces—just enough to do military engineering, say. You needed to build workshops with a certain structure, staff them with workers with expertise, and give them the ability to conduct independent research. You had to revamp Turkish attitudes toward science and technology to make it all possible. When Sultan Selim III opened another School of Military Engineering in 1795, it also encountered strong resistance. In 1803, Seyyid Mustafa, a former student at that institution, recalled being savagely harassed even while studying mathematics: "People screamed at us saying, 'Why do they draw these lines on the paper? What advantage do they think they are getting from them? War is not waged by compass and ruler.' "[1] But Mustafa was less upset by the anger than by the fact that the general ignorance and opposition to his learning made the Ottomans bad citizens, unable to face the empire's current problems, let alone to solve them.

For the moment at least, Mustafa and his peers who studied the "new knowledge" were protected entirely by the authority of the sultan, who was promoting it. In 1808, however, Selim III was assassinated. His successor, Sultan Mahmud II (1785–1839), tried to change the cultural atmosphere. The "Peter the Great" of Turkey, Mahmud II initiated reforms throughout the Ottoman bureaucracy to make it more efficient and bring it more under the sultan's control. These reforms extended throughout Ottoman culture. He introduced Western-style clothing, replacing turbans and robes with fezzes and pants, and military bands that were trained, at Mahmud II's invitation, by the Italian composer Giuseppe Donizetti. The range of reforms eventually dampened ethnic separatism and started to foster a new sense of Ottoman identity among multiethnic and multilingual groups. Mahmud II started the first Ottoman newspaper in the Turkish language, with an occasional science section. Ottoman administrators sent more of the country's citizens to study in Europe, and opened more European-style schools. A significant effect was the beginning of a separation between secular and spiritual realms.

Another effect was the first real debates about science and the "new knowledge" possessed by the Europeans. What was its value? Why was it dangerous to ignore? Was it dangerous to embrace?

This was not a simple debate. What words and concepts could be used to speak about science or the "new knowledge" in an acceptable way? Bacon had also faced this problem in Europe. He had nothing he could point at to show his contemporaries what science was and to illustrate its value, and had cleverly resorted to metaphors and fables from biblical and classical traditions to get across what he meant. The Ottoman rulers had something to point at—but it was developed by people who were not only enemies but heathens.

Two words offered themselves as possibilities for the new knowledge: *ilm* and *fen*. *Ilm* was the Arabic word used for knowledge in its most elevated and venerable sense. It referred to the knowledge possessed by God himself, and as such could become "almost a synonym for Islam."[2] It could also be applied to rigorous fields such as mathematical and medical knowledge, Quran scholarship and jurisprudence, and certain other disciplined fields of study, as well as in some informal contexts. *Fen* referred to the more practical knowledge possessed by scribes and civil servants, surgeons, military engineers, and architects; it's more like the English "know-how" or "art."[3] Was the new knowledge *ilm* or *fen*? The former might be used if you were promoting it as something divine and only accidentally developed first by Westerners, the latter if you wanted to argue that it was entirely secular and could not infringe on religious traditions. Complicating the debate further was that the issue affected social hierarchy in the Ottoman Empire. The members of the Ottoman elite "were not interested in being merely 'men of know-how,' they wanted to be 'men of knowledge,' " writes the sociologist M. Alper Yalçinkaya. Those Ottomans who debated about science, he continued, were not simply arguing about its status, but "carving out a niche for themselves in a social hierarchy."[4]

THE TANZIMAT (1839–1876) AND ITS PUSHBACK

A turning point in Turkey's Westernization came in 1839, when Mahmud II died and was succeeded by his son Abdülmecid, who formalized and extended his father's work. Abdülmecid issued an edict called the Tanzimat, whose measures included European-style systems of currency, legal codes, military training, postal service, and taxation, as well as the abolition of slavery and decriminalization of homosexuality. The Tanzimat reforms involved introduction of several institutions based on European models, such as modern universities (1848) and an Ottoman Academy of Sciences (1851). The Tanzimat also introduced European technology, including the first Turkish telegraph line (1847) and railway (1856).

The Tanzimat reforms also sought to make the new knowledge accessible not just to administrators but also to ordinary people, with the idea that possessing such knowledge made people better citizens who could understand the world and therefore the actions of their rulers. When the Academy was established in 1851, one of its first tasks was to create textbooks in Turkish rather than Arabic, the language of the ancient knowledge.

The Tanzimat reforms reinforced an increasingly glaring cultural contrast between the "new knowledge" of modern and of European (typically French) provenance, and the "old" knowledge, traditionally Turkish and Muslim. This did not involve a simple shift from one kind of knowledge to another. Each was associated with its own set of assumptions, attitudes, and behaviors held by different types of people and even social groups. A catch-all cultural stereotype developed for each: one either did things the traditional way—*alla turca*—or the new way—*alla franca*, the French way.

In *Learned Patriots: Debating Science, State, and Society in the*

Nineteenth-Century Ottoman Empire, Yalçinkaya demonstrates how the clash between adherents of these two kinds of knowledge grew during the Tanzimat from skirmishes between social niches to a wholesale debate about citizenship, morality, and patriotism. Ottoman culture was forced into this debate. Global warming may eventually force ours to have a similar debate, unless our current politicians have the wisdom to initiate it.

Adherents of the old knowledge, typically clerics, regarded promoters of the new as shallow imitators of the European elites. Knowledge that was good enough for traditional Ottomans should be good enough now and need not be replaced. The new knowledge was not spelled out in the Quran nor taught in the *medrese*, the traditional religious-oriented educational institutions. It was niche-learning, not something one could strictly call *ilm*. Those who embraced the new knowledge had lost their dedication to the sultan and the empire, were in danger of losing their morality and religious devotion, and were Muslim in name only.

Promoters of the new knowledge, typically the Ottoman elite, regarded proponents of the old as mystical parasites who were dangerously neglecting the needs of the empire and its people. The new knowledge was not really "European," for the Europeans had simply continued to develop knowledge that had been initiated by Muslim Arabs. Proponents of the new knowledge argued that it is a product of the intelligence that God has given to all humans for their own use—essentially the same arguments that Bacon, Galileo, and Descartes had used against the Church in the West. One could certainly call it *ilm*. If you did not embrace it, you were letting down the empire and its citizens.[5]

During Abdülmecid's reign, proponents of science among the bureaucracy promoted it as Galileo had—as parallel to religious knowledge, but concerned with self-defense and human welfare in this world rather than the next. What ultimately drove the debate, Yalçinkaya writes, was not questions about "the characteristics and consequences of knowledge itself, but the characteristics and attitudes of the people who possessed or lacked knowledge."[6] What was ultimately important was how science affected morality and patriotism.

Abdülmecid died in 1861 and was succeeded by his brother Abdülaziz, who ruled until 1876; their two reigns span the Tanzimat period. Abdülaziz continued the Tanzimat reforms, reestablishing Istanbul University according to the European model in 1861 and expanding the railroads into a large network. In 1867, he became the first Ottoman sultan to visit western Europe. During his reign, the connection between scientific knowledge and the prosperity of the state was propounded as official doctrine; the new knowledge was to be ignored at the state's peril.

Abdülaziz's reign saw the first Turkish language *Journal of Sciences* (1861), published by the Ottoman Society of Science. Its first issue contained an article called "Comparison of Knowledge and Ignorance," which spelled out this official doctrine, linking science (*ilm*), military power, material benefit, usefulness, wisdom, and virtue. While "mindless friends of religion" may think that knowledge harms faith, true believers (of whatever nominal faith) are those who know the mysteries of the universe. The article acknowledged but downplayed the fear that importing European science meant importing secular European values, and its author went out of his way to champion not only the religious value of science but also its service to the state.

The cultural debate about the national, moral, and religious value of science continued to grow in intensity. One reason was increasing awareness of just how different in character the European scientific communities were in spirit and structure from Islamic communities. The forerunner of these European scientific communities was the "Republic of Letters." Participants, including Descartes, wrote letters to each other and circulated these letters within local networks. The members of this republic trusted each other but were capable of questioning each others' claims; they valued sharing knowledge, yet also thought it important that discoverers got credit. But, as science historian David Wootton has described, these local networks grew into an extended one whose character changed. The development of standardized experimental equipment—telescopes, vacuum pumps, and barometers—plus the possibility of regular publication of results, gave birth to a new and more intense scientific culture. Experiments became the principal preoccupation of scientists. The new scientific culture, Wootton writes, was marked by "originality, priority, publication, and what we might call being bomb-proof: in other words, the ability to withstand hostile criticism." The new culture was "innovative, combative, competitive, but at the same time obsessed with accuracy." There's no fundamental reason to think that this is a good way to do science or intellectual life, he continues; "it is simply a practical and effective one if your goal is the acquisition of new knowledge."[7]

This was not how the Ottomans saw themselves. The disparity between the scientific culture of the Europeans and of traditional Muslim scholarship fed a counter-Tanzimat movement among so-called "Young Ottomans." The Young Ottoman movement was a reaction against several trends: the growing secularization of the Ottoman bureaucracy, the increased power of the bureaucracy itself and the elites who ruled it, and the perceived sell-out to European lifestyles. The importation of Western science was a special concern. Though not necessarily opposed to it, the Young Ottomans challenged those who wanted to import Western science to defend its value to the Muslim community, their ability to conform to tradition, and their religious dignity. A striking example involved the Imperial Meteorological Observatory, founded in 1868 on the outskirts of Istanbul (now the Kandilli Observatory). It was criticized in newspapers for not offering the one service that such an observatory could give the Muslim community, namely, the firing of the cannon marking the end of Ramadan.[8]

Among the far-ranging results of the Tanzimat reforms was the rapid growth of the popular press and popular entertainment. The new science became a favorite topic among literate citizens as well as Ottoman administrators, cropping up in opinion pieces and manifestos and also in novels, poems, and plays. The debate, as Yalçinkaya emphasizes, was always more about people than science, inevitably giving rise—as in the West—to sometimes amusing cultural stereotypes. Advocates of the virtues of tradition against the new knowledge were portrayed as authentic and religious, fatherly figures whose deepest thoughts were for the welfare of fellow citizens of the empire—and as parasites who depended on the state's generosity but contributed nothing. Advocates of the new science were portrayed as saviors of the state who were giving new benefits to its citizens—and as imitators of European ways who had abandoned the desire to live spiritually.

Yalçinkaya's descriptions of some of the "villain stereotypes" of Europeanization are amusing. One was the "confused materialist," who was "brilliant in the new sciences but oblivious to the religion and values of his own society." Another was the *şık* or "fop." While the European stereotype of a fop was someone concerned principally with clothes and personal appearance, the Turkish version included a propensity to spout "scientific gibberish uttered in French." A fop was "a Muslim who learned to look, talk, and consume like a European, without any respect for the traditions and religion of Muslim Ottomans or any real knowledge about the topics he discussed—one of which, inevitably, was the benefits of science."[9] Fops appeared as characters in numerous Turkish novels and plays of the time. They were presented as sports fans who know nothing about the game—but were cheering for the visiting team. This didn't necessarily mean that that team was evil, but only that the fan was weird.[10] The fop was the most obvious marker that, in Turkey, "science was not a topic that could be discussed without any presuppositions or implications regarding the proper characteristics of the man of science."[11]

WHO ARE WE?

For the Young Ottomans, then, it was not the practice of science that was suspect, but the man of science. The man of science "should *also* be a religious and thus moral individual," as Yalçinkaya puts it. "A Muslim Ottoman who wished to present himself as a spokesperson for science had to demonstrate that he was indeed a Muslim Ottoman." To counter this challenge, champions of the new knowledge devised arguments to demonstrate that the man of science was an authentic Ottoman citizen, supportive of God and the sultan. In Turkey, unlike in the West, science was defended because of its impact on the character of the scientist and the person who understood science. "The Ottoman man of science," writes Yalçinkaya, "did not claim simply to be learned; he was a morally sound, reliable, and patriotic servant of the Ottoman state."[12] The unintended by-product of this debate, he continues, is that it required not only defending science, but spelling out the nature of the community that was adopting it.

> Ultimately, and very simply, Muslim Ottomans talked about people when they talked about science in the nineteenth century. The entire debate was about what kind of people the Ottomans were (and were not), and what kind of people they should (and should not) become. When Muslim Ottomans talked about science, they asked questions like "What does familiarity with the new sciences transform a person into?" "What does it mean to be an ignorant person?" and "What are the virtues associated with the possession of knowledge?"... They talked about virtue and vice, laziness and industriousness, dependence and self-sufficiency, modesty and arrogance, sincerity and hypocrisy, loyalty and treachery, and contempt and deference. The meaning and bound-

> aries of science (and for that matter, religion) were important questions to ask, but the final answers had to do with people and their qualities.[13]

The central issue in the Ottoman debate about science was a cultural struggle very different from the church-versus-state conflicts that took place in the West. It was, Yalçinkaya remarked, "always about what 'our values' were or, even more fundamentally, who 'we' were." He continued, "The ultimate issue was social order and the key question 'Who are we and who do we want to be?' "

The Tanzimat period ended in 1876, when Abdülaziz was deposed after a series of misfortunes and budget crises. After a brief interim reign, Abdül Hamad II became sultan, the empire's last. Autocratic and paranoid, Hamad II censored the press and rolled back many of the reforms of his predecessors. These actions inspired another resistance movement called the "Young Turks," who demanded constitutional rule and a reduction in the role of religion in government. The Young Turks embraced European science; in 1889 they formed a secret society which (starting in 1895) was called the Committee of Union and Progress, whose name was strongly influenced by Comte's positivism and its slogan, "Order and Progress." Their adherence to European science was so devout that one founding committee member called for the abolition of poetry because it was unscientific. "Religion is the science of the masses," went one of their sayings, "whereas science is the religion of the elite."[14]

Atatürk was born to a Muslim family five years after the end of the Tanzimat period, in Salonica (today Thessaloniki). This ancient city had been modernized during the Tanzimat era, with new European-style buildings and portions of the ancient city wall removed. The momentum for modernization continued in the post-Tanzimat period, with gas street lighting the year he was born (1881), trams (1888), and electricity

(1899). He attended a military academy in Salonika and a military college in Istanbul, in the course of which he became a second-generation Young Turk, joining the Committee and participating in the "Young Turk Revolution" of 1908. This restored the constitutional monarchy and resulted in a multiparty system. The Committee was now able to come out into the open, and as the Party of Union and Progress it took control of Parliament. Atatürk became commander in chief during the Turkish War of Independence (1919–1923), and the year the war ended, the first president of the Republic of Turkey. Atatürk's famous praise of science as the most truthful guide in life, quoted at the beginning of this chapter, was made in 1924, a year after he became president. By then, even the conservative religious factions had largely accepted the inevitable importation of science, and had devoted themselves to reconciling it with Islam. "Unlike other religions, the religion of Islam, which God bestowed upon us, is immune from melting under the bright sun of science," one wrote. "What we need is to prove, by comparing the shining truths of our religion with the lights of science, that these lights were born of these truths."[15]

LEARNING FROM THE TURKISH EXPERIENCE

To a contemporary Westerner, Atatürk's catchphrase about science seems innocuous and self-evident. But it represented the outcome of a grueling century-long cultural debate that is useful now in plotting how to confront science deniers. Always front and center were the following questions, which were posed to all sides of the debate: "What are your values?" "What are your actions going to do to us?" and "Who are we, the community affected by your actions?"

Science deniers often preach different values than they enact, and one strategy can be to force disclosure of their values. North Carolina's legislators, for instance, would surely say that they value the health and welfare of their constituents. But in 2012 they passed House Bill 819—a law prohibiting the use of models of sea level rise to protect people living near the coast from flooding. The law was formulated in response to a report by the science panel of the state's Coastal Resources Commission, which predicted a substantial sea level rise by the end of the century, and reflected fears that the report would harm tourism and property prices. Bills have also been introduced in the US Congress to stop politicians from using science produced by the Department of Energy in policies—evidently to avoid admitting the reality of climate change (so far these bills have failed). During the Trump administration, the Environmental Protection Agency prevented scientific findings from being incorporated into policies designed to protect health and the environment, prompting numerous scientists to resign from the agency.

How, then, do we ensure that politicians genuinely enact the policies they preach?

One way is to ask politicians to sign pledges to show their commitment in favor of or against specific positions about science.[16] This will force them to commit themselves to the values they espouse. Take evolution denial. The president of my university, who is an epidemiologist, likes to say that microbes and viruses are "evolution in motion." Outbreaks of new plagues and viruses thus means that a legislator's belief in evolution, and thus in the value of studying it, is a public health issue. At debates and press conferences, evolution-denying politicians should be asked to sign, or explain why they will not sign, an antievolution pledge: "I pledge that I will not use, nor let my constituents use, any medication whose development depended on evolution or evolutionary theory."

Similar pledges can be crafted to test the sincerity of other science-denying politicians, including antivaccination activists and climate change deniers. The latter should be required to sign (or explain why they will not sign) a pledge to take no action to protect their or their constituents' properties against rising sea levels and other effects of climate change. Donald Trump, for instance, claimed that climate change is "bullshit," "pseudoscience," and "a total hoax." Yet he applied for permission to erect a sea wall to protect one of his golf courses in Ireland from rising seas due to "global warming and its effects."[17] Asking Candidate Trump to pledge that he will take the same actions to protect US citizens from rising seas that he takes to protect his own investments would have exposed that action as not a mere business decision but a betrayal of his constituents.

Another way is to show that science deniers betray their own values. Civilizations have long used scientific methods as a means to discover tools to ward off threats, be they vaccinations to tackle disease, tests to determine the presence of toxins, or ways to create foodstuffs to prevent hunger. Scientific methods discover the levers of nature, one might say. Whether and how to pull these levers is a political decision, but discovering them is a scientific matter. Politicians who try to damage these levers are effectively denying citizens the ability to defend themselves; such politicians are betraying their own values.

Here's an incendiary comparison: US politicians who attack science are like the Islamic State militants who bulldoze archaeological treasures and smash statues. Is such a comparison really over the top? Science is a cornerstone of Western culture, not only to ward off threats but also to achieve social goals. In seeking to destroy those tools, science deniers are like ISIS militants in that they're motivated by higher authority, believe mainstream culture threatens their beliefs, and want to destroy the means by which that mainstream culture survives and flourishes. If anything, ISIS militants are more honest, for they openly admit that their motive is faith and ideology, while Washington's cultural vandals do not. It's disingenuous, prevents honest discussion of the issues, and falsely discredits and damages American institutions. At debates and press conferences, such politicians should be asked: "Explain the moral difference between

ISIS religious extremists who attack cultural treasures and politicians who attack the scientific process." How they respond will reveal much about their values and integrity.

Another strategy is to tell parables involving science denial. A parable, like an Aesopian fable, is a real or fictional story with a built-in, easily graspable moral. It is an effective teaching approach. After all, most people learn more easily through stories than data. *Jaws* is a famous modern example. Fidel Castro—that acerbic critic of anything American—once said he liked that movie because it shows the inevitable consequences of the corruptions of capitalism. He was surely thinking of the scene where oceanographer Matt Hooper, played by the nerdy Richard Dreyfus, realizes that a mangled woman's body is evidence of a shark prowling the waters and tries to persuade the local mayor to close the beaches. The mayor, however, insists that the beaches must stay open because shutting them will be expensive—and the mangled body was probably caused by a boating accident. We know what happens next. When I show this scene of the movie to my students, I tell them that I find it scarier than any bits involving sharks. Another example is Ibsen's play *Enemy of the People*, in which the doctor of a small town whose livelihood depends on its spa discovers that waste from a local tannery is injecting deadly bacteria into the spa's waters. Yet the doctor can't even make himself heard at a town meeting he arranges, and is libeled, accused of conspiracy, and fired.

A final lesson of the Turkish experience is that ensuring the authority of science is not just a matter of focusing on how to communicate—on being clever or witty, or using charismatic people. There are many programs that do so effectively, such as the Alan Alda Center for Communicating Science at my institution, Stony Brook University. Ensuring the authority of science also requires carefully considering the social and historical context. That means the authority of science is not a problem to be fixed, a philosophical puzzle to be solved, nor something that can be guaranteed once and for all. The authority of science has to be constantly justified and reestablished in changing historical conditions by posing and answering questions like: "Who is promoting or denying the science?" "Who benefits?" and "What will its impact on us be?" Restoring and maintaining the authority of science is therefore a problem requiring the skills, not of the sciences, but of the humanities. This is the lesson of the work of Edmund Husserl, whose writings on this are explored in the next chapter.

NINE

EDMUND HUSSERL: CULTURAL CRISIS

We stand at the birth of a new millennium, ready to unlock the mysteries of space, to free the Earth from the miseries of disease, and to harness the energies, industries and technologies of tomorrow.

—Donald Trump, Inaugural Address (2017)

On January 20, 2017, a man was sworn in as the forty-fifth president of the United States who is on record as having declared that global warming is "bullshit" and a "hoax" perpetrated by the Chinese. On the surface, this may not seem like a special cause for alarm. After all, US presidents and political leaders before Donald Trump, from all political parties and persuasions, have downplayed or ignored the relevance of scientific findings for political action. Moreover, Trump was dismissing the findings of only a specific zone of the scientific workshop. He and his supporters presumably still consult engineers to evaluate planned buildings, and rely on the advice of doctors for their health. Furthermore, people may assume that the consequences of such denigration of the scientific workshop are going to be minor, sure that those who do it will never carry out anything unhinged.

But dismissals of science by Trump and other influential American politicians, unlike those of predecessors, are not evasive. These people are confident and proud, and their repudiation of scientific authority is an important part of their appeal to voters. When such people dismiss scientific findings as a "hoax" and "bullshit," they are not mishearing or poorly understanding the findings, nor seeing them as disconnected from social goals—they are straightforwardly rejecting the scientific findings. They are also rejecting the relevance of science and technology for understanding and addressing social issues, assuming that the resulting benefits come free and from out of nowhere. The same person who expressed the desire to "unlock the mysteries of space, to free the Earth from the miseries of disease, and to harness the energies, industries and technologies of tomorrow" dismantled scientific programs including Earth-monitoring satellites, accused scientists of conspiracy, will likely eliminate the Office of the Science Advisor of the Environmental Protection Agency, and appointed lawyers and banking executives in positions calling for scientific expertise who took sledgehammers to scientific procedures and the scientific infrastructure in decisions that affect health and the environment. It is more urgent than ever to incorporate scientific findings into policy decisions about those matters, and it is extremely costly and dangerous to ignore such findings. Yet the president bluntly proclaims that he does not care about those findings, at least in the case of climate change, and his election shows that neither does a large fraction of the US electorate. That has all the makings of a crisis in which the ground underlying key decisions affecting human welfare and safety is no longer solid.

The twentieth-century philosopher Edmund Husserl (1859–1938) understood the dangers of such a situation, having lived through a similar cultural crisis himself. Unlike Weber and like Comte, Husserl thought the crisis might be overcome. But Husserl realized that overcoming it would be harder than even Comte, with his utopian leadership ideas, foresaw. Husserl came to believe that the very ground on which humans were standing when they made rational decisions had been undermined, and partly by the impact of science itself. That ground had to be fixed before the crisis could be confronted.

Edmund Husserl (1859-1938).

EXPLORER OF THE LIFEWORLD

Born in 1859, Husserl was the second of four children in the family of a Jewish clothing merchant who lived in a town in Moravia, now part of the Czech Republic, and at that time a region in the Austro-Hungarian Empire. It was a temporary period of freedom and fortune for Jewish communities in the empire, who had been formally awarded equal rights under the law by the emperor. Though small, these communities had an outsized impact on German-speaking culture, arts, and sciences. The composer Gustav Mahler and the psychologist Sigmund Freud were born in Moravia during that time, and many Moravian Jews nurtured hopes of becoming fully participating members of mainstream German life. They were sometimes described as "Jewish families of German culture."

Instead of sending Husserl to the local Jewish technical school, his father sent him away to Vienna when he was only ten years old to receive a classical education. Husserl acquired enough of the ideals of mainstream culture that, in his late twenties, he converted to Christianity and read the New Testament regularly. He married Malvine Steinschneider, another Jewish convert from his hometown, and the couple soon had three children: a daughter Elizabeth, and two sons, Gerhart and Wolfgang. In every respect, Husserl and his family were culturally fully assimilated, and poised to participate fully in the social, cultural, and professional life of the German-speaking world on level ground.

But that ground, it would turn out, was more unstable than it seemed.

Husserl was the first of those we have encountered in this book who was a professional philosopher. He was a university professor who, to the virtual exclusion of all else, spent his entire life examining ideas and communicating the results to students. He also shared the conviction of professional philosophers that doing so would have a beneficial impact on human activities, and perhaps human life itself. He once described himself as akin to an explorer who opens up an unknown land so that others, using more conventional habits and techniques, can cultivate it. But the stormy politics of the first half of the twentieth century made his explorations more turbulent, and more relevant to contemporary historical developments, than even he anticipated.

Husserl studied science before receiving his doctorate in mathematics from Vienna in 1882, and as a philosopher he explored the foundations of mathematics and science. He was convinced that one could discover these foundations by paying careful attention, not to psychology nor to pure logic, but rather to how humans experience the world prior to the use of analytic methods, strategies, and calculations. This insight became the basis of his first major work, the *Logical Investigations* (1900–1901), which examined the foundations of logic and mathematics. The promise of this brilliant work led David Hilbert, one of the greatest mathematicians of all time, to invite Husserl to Göttingen, a famed center of mathematical studies in Germany's heartland.

Thanks to Hilbert, Göttingen's Mathematical Institute was becoming one of the world's leading centers of mathematical studies. Its members, who included several Jews, were renowned for analyzing mathematical variances and invariances in a novel way. Mathematicians had studied geometrical objects such as conic sections since ancient times, discovering that you can get a circle, ellipse, parabola, hyperbola, or even a point or line by slicing a cone with a plane. These geometrical figures are, as it were, different ways that a plane can "see" one and the same cone. Many geometrical figures are thus unified by being described as different operations performed on a single object. The Göttingen mathematicians pursued this insight throughout mathematics. They sought to unify infinite numbers in already known mathematics by describing them as the result of applying different operations to the same mathematical object. This approach supported the idea that mathematics had a fundamental coherence and structure that neither springs from the human mind nor is in specific mathematical objects and formulas, but can be found by patient study. This allowed the Göttingen mathematicians to describe new mathematical objects and relationships among mathematical objects.[1]

In Göttingen, Husserl effectively extended this approach to perception. When you perceive a desk, you never grasp it all at once; some parts are out of sight wherever you stand. The desk, like the cone, has profiles that flow into one another as you walk around it, open its drawers, see it in different lighting, and so forth. You can shut and reopen your eyes, or walk out of the room and return, and still grasp the same desk. It's not a chaotic blur but *one* desk: built into your perception is the assumption that it has infinite horizons and profiles, some of which can be unexpected and surprise you when you encounter them. This allows you to grasp the desk as an invariance amid an ever-varying set of perceptions fulfilled in different ways. In Husserl's technical vocabulary, you "intend" the same desk, which "transcends" the different experiential acts by which you grasp it.

As for cones and desks, so for everything else. Things you can intend as transcending your ever-changing perceptions include not only physical objects but colors, emotions, ideas, and events. Otherwise, you couldn't speak of the same shade of red in different lighting conditions, or argue with other people about the meaning of ideas like equality and courage. Furthermore, you can intend these things not just by perceiving but by other means—remembering last week's meeting, for instance, or imagining next month's wedding. The flow of experience is not a chaotic blur; things appear in that flow as apart from it because your experience is structured. The fact that some of your perceptions of an object can surprise you is not a flaw in perception, but part of what convinces you that objects are inexhaustible and "out there," independent of your perception. This structured flow of experience is why you experience *one* world along with everyone else, rather than everyone experiencing different worlds. The experienced world—Husserl will eventually call it the "lifeworld"—has a fundamental coherence and structure that can be brought to light by patient study.

A little over a century earlier, the philosopher Immanuel Kant had come to a similar conclusion, arguing that philosophers could deduce the abstract forms of such structures by asking what made experience possible, in what Kant called a "transcendental" method. Husserl, however, argued that we can find these structures not abstractly and deductively but directly and immediately, by paying careful attention to our experience—to how phenomena present themselves to us in the most original way. Using such a descriptive method, Husserl thought, allowed a deeper, richer, and more complete characterization of the world than Kant's transcendental method. Husserl called his approach *phenomenology*.

Doing science is only *one* way in which humans experience the world, and not the default setting. Humans engage the world in many different ways. They seek wealth, fame, pleasure, companionship, and other things—and as children, adolescents, parents, merchants, athletes, and administrators. As Descartes is at pains to relate in the *Discourse*, humans are not automatic information absorbers; they must be trained to approach the world as he does. Phenomenology describes how different activities, including science, spring from the ground of the world prior to such training, and how these activities make such training possible.

CONTROVERSIES

Husserl spent his years at Göttingen developing and extending phenomenology. In 1910, he helped found *Logos*, a journal that would be devoted to his new approach. For the first issue he wrote a manifesto-like essay, "Philosophy as a Rigorous Science." It was shockingly ambitious. His aim, he wrote, was "boldly and honestly" to expose "the unscientific character of all previous philosophy" and to describe a "revolution" in philosophy. What Descartes had begun, phenomenology gives us the tools to finish; we must set stone "upon stone, each as solid as the other" to rebuild philosophy "from the ground up."

By "all previous philosophy," Husserl had two specific targets in mind. One was called naturalism; the other, a reaction to the first, was known as historicism. The first essentially viewed the world from the perspective of the workshop, the second essentially viewed the workshop from the perspective of the world.

Naturalism is the view that reality is what turns up in a scientific workshop. Because science has such great success treating nature as a network of forces and causes, scientists and others are tempted to use its tools to understand everything that way. The naturalist adopts the slogan "Make and test theories!" and applies it to everything from art, beauty, and consciousness to ethics and values. The naturalist regards doing anything else to be as crude and unsophisticated as pre-Galilean natural science. But naturalism easily turns into a "scientific fanaticism" that scorns knowledge unable to be demonstrated with scientific exactitude. Philosophy, Husserl thought, can become a rigorous science not by making and testing theories, but by returning to the ground floor of experience and examining how the world presents itself to consciousness.

Historicism, a reaction to naturalism, views reality only in terms of what is outside the scientific workshop. After all, that is where human beings, even scientists, live. That world is shaped by values, goals, morals, and spiritual behavior, and its character changes from age to age. While naturalism seeks to use the workshop to reduce all worldly phenomena to a causal network, historicism tries to see all human phenomena, including scientific workshops and their findings, as manifestations of ever-changing worldviews, inherited and transformed from one generation to another. The historicist "product" is not theories, but rather descriptions and comparisons of general perspectives on life. While historicists are able to examine in an undistorted way many things that naturalists do not, such as features of art and culture, in the end historicists leave you with statements by experienced and wise people on how you should live.

Here's an offbeat illustration: to surfers, writes William Finnegan in *Barbarian Days*, his Pulitzer Prize–winning memoir, sizing up waves is a topic of "perennial dispute." Some call waves waist-high, head-high, or overhead. Others judge waves by how many refrigerators-stacked-on-each-other they are. Conventional measures are such nonsense, Finnegan says, that a surfer who spoke of a nine- or thirteen-foot wave "would be laughed off the beach." What matters to surfers about surf cannot be captured in a specific measurement. "Two waves of the same height may differ enormously in their volume, in their ferocity." Attitude also matters: macho surfers underestimate wave size. "Big waves are not measured in feet," runs the surfers' maxim, "but in increments of fear." The same can be said for sailors, fishermen, swimmers, and others who are intimately connected with waves: how one encounters waves guides how one measures them.[2]

The trouble with naturalists is that they would try to use the tools of science to find objective measures of things like waves apart from how surfers, sailors, and swimmers experience them. That would isolate the phenomena from life as humans live it. The trouble with historicists is that they would seek to describe the different outlooks of surfers, sailors, and swimmers on waves, isolating the phenomena from what can be known scientifically about them. Both separate abstractions from life and thereby cut themselves off from the ability to understand the world deeply enough.

In his bold manifesto, Husserl claimed phenomenology could do it all. Using phenomenological as well as scientific tools—sticking to the ground zero of life by relying on firsthand experience—he could describe how humans encounter such things as waves in different ways. Phenomenology's slogan is therefore "Away with empty word analyses! We must question the things themselves!" Only then can we redo philosophy "from the ground up."

HONING PHENOMENOLOGY

Almost immediately, historical events brought phenomenology into conflict with prevailing intellectual tendencies. The first was with naturalism. During Husserl's years in Göttingen (1901–1916), science and technology enjoyed high cultural prestige as unalloyed blessings for humanity throughout Germany and all Europe. Science promised an orderly, stable, and mechanical picture of the world, thanks to the use of such relatively new discoveries as the fourth dimension and X-rays to discover structures of reality just beyond our perception. Technological advances resulted in increased material comforts such as electricity and radio, automobiles and aircraft. In 1908, the flight of one of the first German Zeppelins attracted tens of thousands of spectators; when it was damaged after an emergency landing, a wave of donations from the enthusiastic public raised enough money, in twenty-four hours, to rebuild it. The Futurist art movement celebrated technology and its power, its Italian exponent Filippo Marinetti famously declaring in his "Futurist Manifesto" of 1909 that "a racing car . . . is more beautiful than the *Victory of Samothrace*."

During this period phenomenology was a counterforce to naturalism. Husserl elaborated key concepts of phenomenology that demonstrate that there are more elements to the human experience of the world than show up as objects in the workshop. Workers in the workshop adopt what Husserl called the natural attitude, or how humans view the world in ordinary living. We take the world for granted without noticing the presuppositions—the bundles of horizons of profiles, for instance, or the flow of time—which make a world appear rather than a blur. These presuppositions only emerge through *bracketing*, a concept inspired by Husserl's mathematical training, in which we set aside our interests in the world (why we want to use the desk, say) to examine how we experience it.

Then, in 1914, Germany entered the Great War—so called, rather than World War I, because this first global war did not yet have a sequel. As the conflict dragged on and its horrors became more severe, Germany's lack of preparations for a lengthy engagement took an increasing toll on the lives of its citizens. Husserl himself suffered personally. His younger son Wolfgang was killed at Verdun, his elder son Gerhart was badly wounded and lost an eye; his daughter Elizabeth worked in a field hospital. Deeply shaken, Husserl had to give up working for a year. He changed universities, and in 1916 moved from Göttingen to Freiburg. Until the war's end he, like his compatriots, was sure that God was on Germany's side, and that his nation would triumph.

Until the fall of 1918, in fact, most Germans still believed that they would win the war one way or another. Many were only convinced that Germany had indeed lost when they saw foreign soldiers marching unhindered on their soil. The stunning defeat and its aftermath sharply changed Germany's social and political climate. The humiliating and oppressive reparations that the Allied Powers imposed on Germany virtually destroyed its economy and fostered the rise of nationalism. Meanwhile, the promise of science and technology now seemed a bad joke. It had not brought peace and prosperity, but had magnified the war's horrors, giving the combatants newer and deadlier tools—machine guns, flame throwers, tanks, submarines, aircraft, poison gas, and more. These multiplied the casualty rate among soldiers and civilians more than in previous wars. After the war, many countries experienced a cultural backlash against the embrace of scientific rationality. Art movements, such as the Dadaists and Surrealists, now mocked and satirized technology and the culture of rationality (an example is André Breton's "Surrealist Manifesto" of 1924), which they saw as partly instigating the war.

The result, as the historian Paul Forman has written, was the emergence of "a neo-romantic, existentialist 'philosophy of life,' reveling in crises and characterized by antagonism toward analytical rationality generally and toward the exact sciences and their technical applications particularly."[3] The author who epitomized this spirit was the German historian Oswald Spengler, whose best-selling book *The Decline of the West* argued that civilizations are like organisms in that they—and all their cultural practices, including science—follow a developmental pattern in which they rise, develop, and decline. Thinkers all over Europe began to regard the turmoil they were experiencing as not just a temporary political crisis, but an epochal moment in the history of civilization.

Now Husserl and other phenomenologists found themselves waging war against another enemy, historicism. In Freiburg, where he would remain and work even past his retirement in 1928, he acquired many students and followers who helped him in this task, such as Martin Heidegger (who dedicated his key work *Being and Time* to Husserl), Karl Jaspers (Heidegger and Jaspers first met at Husserl's sixty-first birthday party in 1920), Hannah Arendt, and Edith Stein (who served as Husserl's assistant for two years). One of Husserl's most famous and influential series of lectures, edited by Heidegger, concerned the nature of time. The experience of time has certain basic features, he argued. All experience has a unity across an extended flow, from remembering and planning to listening and perceiving. In this experience, time is not experienced as one point succeeding another, though this is how scientists may define it. It is an asymmetrical flow characterized by the retention of what one has just experienced and the protention, or anticipation, of what may come next. We hear a melody, for instance, not as a series of individual notes that we then string together, but as already temporally extended; we retain notes already heard and anticipate those to come as part of the melody. Later, I can speak of some musical thing happening at seven or ninety-two seconds into the melody as if that were a particular moment. Yet that's only because I've already heard the melody in the first place as a temporal flow.

SHAKY GROUND

But German culture, whose ground had been seemingly solid, was becoming ever more precarious. Developments that included rising inflation and the French occupation of Germany's wealthy Ruhr Valley to siphon off its resources (1923) led to political unrest. From 1924 to 1929 Germany experienced a period of relative stability, but the "Golden 1920s" abruptly ended with the stock market crash of 1929, which brought unemployment and political turmoil. Throughout the 1920s and early 1930s, the National Socialists exploited the growing cultural anxieties. Hitler, who had led the party from 1921, managed to become chancellor in January 1933, bringing the National Socialists to power. Still working feverishly at his phenomenological research, Husserl had not been politically active. No doubt he assumed that because he had converted to Christianity he was protected from the anti-Semitic Nazi rhetoric around him.

He was not. Early in April 1933, the Nazis promulgated a law forbidding non-Aryans from holding state posts. The term non-Aryans included those of Jewish background; state posts included university positions. Meanwhile, Heidegger became rector of the University of Freiburg and, as expected for someone in that position, joined the Nazi Party. Once there, Heidegger carried out certain Nazi decrees, and signed the order that sent Husserl into a forced leave of absence. Shocked, Husserl successfully pleaded for exemption because of his children's service in World War I.

Husserl's surviving son Gerhart was fired from his university position. Many of Husserl's Jewish students also suffered. In 1933, Arendt was arrested and imprisoned by the Gestapo; when she was released, she fled to Paris. Edith Stein, who had converted to Christianity in 1922 and had been teaching at Catholic institutions ever since, was forced to quit teaching. Fearing for her life, she entered a Carmelite convent for protection.

In the summer of 1934, Husserl received an invitation from the International Congress on Philosophy to speak about "the mission of philosophy in our time" at its upcoming Prague meeting. The invitation jolted Husserl into a new research project to which he would devote the rest of his life. The next year, he gave a lecture in Vienna on "Philosophy in the Crisis of European Mankind," followed by lectures in Prague. A journal in Belgrade agreed to publish his evolving treatise in installments as he completed them.

Husserl's situation under the Nazis deteriorated. Conversions and a family's sacrifices to the first world war effort were no longer grounds for exemption to laws against non-Aryans. In 1935, he lost the right to teach and his German citizenship was soon revoked; Husserl was now officially regarded as non-German. His name was stricken from the list of faculty at Freiburg, he was denied permission to attend international philosophy conferences, and he was forbidden to speak or publish in Germany. The Nazis attacked his idea of a universal rationality as absurd, since it would include Jews and Negroes.[4] He and his wife were forced to leave their home for elsewhere in Germany.

Many prominent Jewish scholars left Germany. Husserl was content to stay and work intensely on his project to the exclusion of everything else. He declined an offer of a position at the University of Southern California.

Husserl's project turned into a multipart treatise about the mission of philosophy. The Belgrade journal published the first two parts in 1936. Husserl had begun a third and had two more planned when, in 1937, he fell ill of terminal cancer and had to stop. What he wrote is hurried, patchy, and difficult reading. "Their literary quality leaves something to be desired," dryly writes his translator, David Carr.[5] The full work was published years after Husserl's death—in German in 1954, in English in 1970—as *The Crisis of European Sciences and Transcendental Phenomenology*. In it, Husserl tried to explain the cultural vacuousness that had seemed to descend on his country and elsewhere in Europe.

THE *CRISIS*

Though repetitive and convoluted, Husserl's last work contains deep insights into the relation between the workshop and the world. It describes how the modern scientific workshop sprang from the lifeworld, and its particular strengths: that it is a collective activity, that it is technical and abstract, and that it is ongoing. But Husserl also indicates that, in particular cultural contexts, these very strengths make it possible for people to suspect and even repudiate the workshop and its products.

That can happen, Husserl concluded, because the sciences only answer technical and factual questions and have nothing to say about the meaning of human existence. Furthermore (and in a way that recalls Vico's barbarism of reflection), the products of science can be used without understanding what is entailed in producing them.

To show this, Husserl essentially fashioned the *Crisis* as a story, though it's a rather tangled one. I will condense and simplify it.

Prior to modern science, Husserl says, science was conceived as a comprehensive study of reason that encompassed both the human and natural worlds and saw them as intertwined. Leonardo da Vinci's painting projects, for instance, caused him to make careful and elaborate studies of muscles and bones, which then improved his painting. In the early modern period, however, Europeans began to develop a new conception of science, building workshops to manufacture scientific tools. These workshops provided human beings with maps that, as Bacon had envisioned, allowed humans to navigate the world better. The new "Galilean" science, as Husserl called it, portrayed the universe as a vast geometrical space whose shapes and relationships are mathematically determinable. But it also had the key undesirable consequence of fostering the belief that the maps provided by science are reality itself. The reality in which we are bathed when we wake up in the morning, share our family life, make friendships, play and work, and hope and fear, fades into the background as something subjective in comparison to the objective maps provided by the workshops. This distortion of how we view reality is behind the current cultural and spiritual crisis.

Philosophy ought, if any discipline can, to be able to come to the rescue and provide the right perspective on reality—but it, too, has been distorted. The humanities are too feeble to help, because they have been cowed by the sciences into abstaining from questions of value or purpose in the name of objectivity. In the *Crisis*, Husserl mentions two philosophical movements that have gone off the rails in different directions. One is positivism, a virulent form of naturalism, while the other is existence philosophy, a more histrionic variant of historicism that would soon be called existentialism.

Positivism had grown far away from Comte's original vision, which had involved a flexible method, sought order and progress, and kept people on their toes by retelling the Great Law of intellectual evolution. Positivists were now different. Post-Comtean or nouveau positivism viewed the Great Story as unnecessary now that the scientific stage was nigh. It considered the scientific method to be single and fully developed; scientific activity to be the work of a contextless, abstract mind; and scientific knowledge to be divorced from the historical, political, and social milieu in which it originated. The real was what showed up in the workshop, and everything else was either to be understood in terms of the workshop, or seen as obstructive and potentially dangerous. Positivists don't understand the current cultural crisis, Husserl says. They are like people living in a dysfunctional building who blame people outside the building.

Husserl thought he saw the beginnings of a reaction in *existence philosophy*, concerned principally with things of immediate relevance to the lifeworld. Some of Husserl's own students and coworkers had contributed to such philosophies, including Heidegger and Karl Jaspers, a professor in Heidelberg. Existence philosophy begins with the assumption that the world as given, outside the workshop, is culturally and socially meaningful. Existence philosophy tends to regard workshop thinking as a threat to understanding the lifeworld, which is shaped by human practices and therefore not theoretical. Existence philosophers, in Husserl's eyes, were like people in a dysfunctional building who blame the building's carpenters, electricians, or plumbers.

In his "Rigorous Science" essay and other writings, Husserl assumed that bracketing and other techniques could rid philosophy of the errors and prejudices of such movements and purify its vision, so that philosophers could then stand on level ground and see the world right. In part thanks to Heidegger's writings, Husserl now realized that was not sufficient, for the ground was uneven. In becoming members of Western culture, we have not only acquired facts, truths, and other information, but also values and goals that shape our perceptions of the world even before we start bracketing. Specifically, Galilean science has corrupted our ability to understand the world even before we intend and bracket phenomena to study them. Galilean thinking has led us in the West to instinctively mathematize experience, which is a special, even artificial way of thinking. First we have to undo this corruption. How? By telling the story of how this happened. To appeal to the above analogy, the most thorough way to understand a dysfunctional building in preparation for fixing it is to trace how it came into being.

The story Husserl then proceeds to tell is like Comte's Great Story in that it is a tale of intellectual development. But it has even stronger resemblances to Vico's Great Story, whose protagonist—humanity—thinks that it is progressing when in fact it is following a path to decline.

This part of the story begins in Section 9 of the *Crisis*, in which Husserl attempts to reconstruct Galileo's train of thought in his remarkable mathematization of nature, embodied in his famous image of the book of nature written in mathematical figures. Since ancient times, Husserl points out, geometry and mathematics in general have been applied to the world in practical applications like accounting, carpentry, navigation, and surveying, whose practitioners act "as if" the world were implanted with ideal shapes and mathematical formulae, amenable to infinitely precise measurements and calculations. Geometry and math provided maps that connected abstractions with real and concrete phenomena, enabling humans to navigate the world better.

These maps were so successful, Husserl writes, that they misled some of those practitioners into taking these maps as reality itself. Some aspects of reality, such as colors, warmth, and weight, were not directly mathematizable, but these "secondary" qualities could be indirectly mathematized through things like wavelengths and number-labeling. Other elements of reality that were scarcely mathematizable at all, such as values and goals, faded into the background. The new move had many powerful advantages, including the promise of certainty, infinite precision, and generality. Mathematizing physics meant that it is not just *this* body falling, but *a* body falling; not *an* attraction between *these* two bodies but *all* bodies. At last, science had become universal.

It was a turning point in Western civilization. Until then, math and science were seen as providing knowledge *about* reality; now they were reconceived as determining *reality itself.* "As if" became "It is!" and "to be" became "to be measurable." Galileo's work opened up a powerful new path for the West—aiming for universal knowledge—but one that was also treacherous. Galileo, whom Husserl calls a "revealing and concealing genius," surely did not see this; he was working out his method intuitively. There is an embedded pragmatism in his procedure and in the scientific method. They both judge science in terms of its results, in what it yields, not its processes. The scientific workshop thrusts its processes into the background and is forgetful of them. You therefore can't criticize science by saying that it's abstract. Of course it is—that's what makes it effective! But it looks as though you have to choose between the abstract world of the workshop and the world outside and its values. If you take one, you have to neglect the other.

Something strange has happened. Workshop findings can dazzle human beings, making the surrounding world—the one in which we make friendships, breathe and suffer, hope and fear—appear to recede in importance. This fading in importance of the humanities, Husserl thought, was behind the cultural crisis of his time, most notably in the form of Nazism. But the effect of weakened humanities can also be a loss of the understanding of what science is all about. The deep structure of the workshop and its tie to the lifeworld is lost. You think you can pick and choose your scientific findings. You think scientific findings are opinions. You can set out to "unlock the mysteries of space, to free the Earth from the miseries of disease, and to harness the energies, industries and technologies of tomorrow," as Trump put it, without supporting the scientific infrastructure that made it possible.

Is there a way to recover this deep structure of the workshop? Like Comte and Vico, Husserl thought it was only possible by following the story of how it all happened. But this is something scientists themselves—"unphilosophical experts," Husserl calls them—are unprepared to do. Scientists do not even experience the need to tell that story, for they achieve their successes by focusing on their work and looking away from the life-world. Scientists are like the operators of a machine that is extremely useful for some tasks, but controlling the crisis depends on individuals who are able to reflect on the possibility and necessity of such machines. The important story therefore turns out to be the very one that Husserl is in the process of telling.

But how has Galilean science distorted the world outside the workshop?

For one thing, the tools, methods, and imagery of the workshop have spread to the world. These things can be handed down and used without repeating or understanding the original discovery. Computers, cellphones, and microwaves shape the worlds of today's children. But it's more than a matter of tools. Humans now have a quantitative understanding of the world thanks to practical experience with instruments such as digital clocks, speedometers, scales, and thermometers. More ominously, though, what humans readily download from the world thanks to Galilean science is a stance toward reality itself. It's not just that we learn to explain the world scientifically, but that we begin to interpret ourselves, others, and our relations with the world in scientific terms.

The world seems "obviously" divided into scientific and nonscientific, objective and subjective, theoretical and practical spheres. This has a huge cultural cost. It denigrates—maybe not explicitly, but implicitly—most aspects of art, culture, and spiritual life. It encourages those who value those aspects to reject science as a detached and elitist practice—one factor at work in contemporary science denial. Husserl was forced to conclude that historicism is partly right: the way we encounter the world is shaped by contingent historical factors that mold our outlook at the deepest levels. Before turning things into objects inside or outside the workshop, I already make presuppositions about reality—about what the important things are. Husserl hoped that his retelling of the story of how this all happened would be the first step toward breaking the spell.

The storytelling will focus our attention on "who" we Westerners have become thanks to our intellectual inheritance.

Comte thought that thinking scientifically was fine as long as you don't forget about how it arose out of human life and concerns. Vico thought it was fine if supplemented by the humanities. Husserl now thought that science has dazzled us into being unable to do even that. You have to tell a deeper story than Comte and Vico told, one that describes what happened not naturalistically but phenomenologically; that is, one that pays attention to how Galilean science has affected the ways we pay attention to the world, threatening to saddle it with a cultural emptiness.

A HUSSERLIAN NEW ATLANTIS

One can imagine a Husserlian New Atlantis that would draw on natural curiosity to foster scientific curiosity and exploration of the world. It might be full of exhibit halls with things like tops, pendulums, and telescopes that give a palpable presence to vastness in small things. It would engage citizens across the planet to participate in "civic webs," or projects that maximize interactions between the inhabitants and nature as well as each other on shared problems, with such interactions also fostering appreciation of the value of the webs themselves in addressing such problems. Such civic webs would teach practical wisdom: how to look for what you need when you don't yet have it.

This is not as difficult as it sounds. Nearly everyone has a proto-scientific attitude—a sense of how to inquire, test, and discover—that can be nurtured. The mystery is how it can be misunderstood and even lost—which is the lesson of the Husserlian Great Story. The aim of a Husserlian New Atlantis would not be to change people into scientists, but rather to encourage their appreciation of its value in understanding and coping with the world. Science is fated to have a dual destiny in any community or civilization. It is a rigorous discipline with its own norms and vitality independent of the general culture—but it is also historically and practically important to that general culture. The special aim of this New Atlantis would be to make sure that last insight is not lost, which could be done by retelling the Husserlian Great Story of how science can cause us to lose sight of its own moorings.

═══

HUSSERL DIED IN 1938, three weeks after his seventy-ninth birthday. Only one member from his philosophy faculty attended the funeral two days later. Not even Heidegger, who had succeeded Husserl in his position after the latter's retirement, showed up; Heidegger later claimed that he had been in bed with the flu. Realizing that Husserl's remaining manuscripts were in danger, a Belgian priest who had been studying Husserl's writings managed to convince his country's diplomatic courier to smuggle them to Belgium, and arranged for Husserl's widow Malvine to hide in a Belgian convent for the rest of the war.

Some of his students were less fortunate. Hannah Arendt, who had moved to Paris in 1933, was stripped of her citizenship; following the Nazi occupation of France in 1940, she was shipped to an internment camp before managing to flee to the United States the next year. Husserl's former assistant Edith Stein was not safe in her Carmelite convent in Germany. In 1940, she moved to another Carmelite convent in the Netherlands for protection. In 1942, after the German invasion of the Netherlands, she was sent to Auschwitz. She was gassed two days after her arrival.

IV

The philosopher Hannah Arendt's book *The Origins of Totalitarianism* (1951), which first explained to Americans the background of what would soon be known as the Holocaust, brought her to public attention. But in 1963 Arendt became a household name after a five-part series of articles that she wrote for the *New Yorker* magazine on the trial of the Nazi war criminal Adolf Eichmann was published as a book, *Eichmann in Jerusalem*. It was there that she introduced the controversial phrase "the banality of evil."

In her book on Eichmann and her responses to often vicious criticisms of it, Arendt confronted how an unskilled person like Eichmann could have acquired the ability to play an instrumental role in the mass deportation, incarceration, and extermination of people in one of the most horrific acts in the history of humanity. Part of the answer had to do with the power that science and technology, born of the ideas of Bacon, Galileo, and Descartes, delivered to the modern world. Another part had to do with the accompanying uncertainties, displacements, and specialization that science and technology had also brought, in insights related to those of Vico, Shelley, and Comte. Still another part had to do with bureaucracy and the breakdown of authority that Weber described, the dangers of "acting into nature" that Shelley depicted, and the cultural vacuousness to which Husserl pointed. Above all, like the participants in the Ottoman debate described in Chapter Eight, Arendt was interested first and foremost in the impact of the powerful new forms of science and technology on how human beings lived and who they were.

Arendt's work allows for a remarkable synthesis of the implications for science denial of the previous writers. It provides a new and timely framework for understanding the dynamics, and goes beyond the disbelief, diatribe, and easy moralism with which the subject is usually discussed.

TEN

HANNAH ARENDT: ACTION

To accept one's past—one's history—is not the same thing as drowning in it; it is learning how to use it.

—James Baldwin

When inducted into the American Academy of Arts and Letters in 1964, Hannah Arendt (1906–1975) said the "extraordinary events of the century" had turned her into a writer by "accident." How morbidly fortunate for the world. Born to a Jewish family in Germany, she lived through the crisis that her teacher Husserl wrote about, and was stateless for eighteen years after fleeing when the Nazis took over in 1933.

In 1961, Arendt covered the trial of Adolf Eichmann for the *New Yorker* magazine in a series of articles published in book form as *Eichmann in Jerusalem*. Eichmann was on trial for war crimes for his role in Germany's mass extermination of Jews and others during World War II, one of the largest and most heinous crimes against humanity in history. Yet he was not personally accomplished or capable. He owed his first real job to family connections, he pontificated in recycling stock phrases and clichés, he borrowed and then mangled ideas from others, and he denied firmly established facts. The special political conditions of Germany at that time, however—the cultural vacuousness described by Husserl—gave great power to braggarts like Eichmann who coveted media coverage, had no interest in truth, and polarized complex situations by casting them as issues of "us" versus "them." How could this possibly happen? Arendt tried to answer.

ON THE MARGINS OF KING'S MOUNTAIN

It is hard not to digress when writing about Arendt, and discuss her remarkable youth, friendships, lovers (who included Martin Heidegger), arrest, internment, flight to America, and various battles. Her work would be easier to write about had she been a traditional or systematic philosopher. Her writings have deep philosophical foundations, but she was suspicious of applying theories: where did these theories come from, she would ask, and why *these* theories and not others? Arendt had instead a phenomenological instinct. She did not theorize, but provided patient, lengthy descriptions, informed by her reflective judgment, of the specific situations she was living through. "Thinking without a banister," she called it.

Hannah Arendt (1906–1975).

Königsberg, where Arendt spent her childhood, was a Prussian city built around a central high fortress; its name means literally "King's Mountain." It is famous as the birthplace of the philosopher Immanuel

Kant (1724–1804), where he lived and died without traveling far outside the city limits, and where he established one of the greatest and most comprehensive of philosophical systems. Kant's system, a fully elaborated universal ideal of rationality and morality, embodied Enlightenment ideals. A pillar of the philosophical canon, Kant's system needed—and still needs—to be studied by every philosopher. It had little explicit room for diversity, however. While respecting certain Jews as individuals, for instance, Kant found their religion was not cosmopolitan enough to fit the ideals of the Enlightenment, and wrote that the Jews would ultimately have to give up their outmoded traditions and embrace Christianity in what he called the "euthanasia of Judaism."[1]

Arendt had a different experience than Kant of the King's Mountain. From an assimilated Jewish family, still she never knew the sense of a culturally firm ground that even Husserl had briefly experienced. Her family lived on the community's margins and experienced numerous displacements. Anti-Jewish legislation had forced her ancestors to flee Russia, and they joined a community of assimilated Jews in Königsberg. For a while her father had a job in the nearby city of Hannover, where Hannah was born. But in 1909 he grew too ill to work and returned the family to Königsberg, where he had connections and where he lived another few years.

Arendt's parents aimed to raise her to become a fully participating member of German cultural life. Her family was so unreligious that, she recalled, "I did not know [at first] from my family that I was Jewish." The news arrived early in childhood via anti-Semitic taunts; "after that I was, so to speak, 'enlightened.' "[2] Her mother then laid down strict rules for coping. If exposed to taunts in school, Hannah was to "get up immediately, leave the classroom, come home, and report everything exactly," after which her mother would lodge a formal complaint by registered mail and Hannah would get the day off from school. If exposed to taunts out of school, she had to stand up for herself and not talk about it at home. These measures, she said later—observe, report, and describe—helped prepare her to be a writer and prevented her from having her soul "poisoned" by anti-Semitism.

In 1914, a year after her father died, the First World War broke out and her mother took her to Berlin ahead of the advancing Russian army; ten weeks later, after the Russian progress was halted, they returned. Hannah's mother was now poor and the postwar years were hard. In 1920, Hannah's mother married a widower with two daughters of his own; moving in with them displaced Hannah once more. But by then she had built a protective shell, and had no trouble sticking up for herself even when there were repercussions. As a teenager she was expelled from school after organizing a boycott of an offensive teacher's classes. She transferred to another school from which, thanks to her independence and intelligence, she graduated a year ahead of her former schoolmates.

German students who passed the university entrance exam could attend whatever university they wanted, and generally chose one with an influential or charismatic professor in their field. Arendt's then-boyfriend had moved to Marburg to study with a philosophy professor named Martin Heidegger. Heidegger sounded interesting. Arendt chose Marburg.

LOVER OF THE HIDDEN KING

Years later, for Heidegger's eightieth birthday, Arendt wrote about him in the *New York Review of Books*. In 1924, she said, Heidegger's fame among students did not rest on anything tangible; he had published no books, no influential articles. His reputation rested on reports of what happened in his classes. Like many philosophy professors, Heidegger devoted his classes to figures from the history of philosophy. Unlike others, Heidegger did not provide canned summaries, nor an "ism" based on a fundamental principle. Instead, he took the stage of the lecture hall each day to engage in an Olympic wrestling match with some past philosopher—a struggle always with unexpected and surprising moves—the prize being the ability to think freshly. The word on the academic streets was that Heidegger was a rebel: "Thinking has come to life again; the cultural treasures of the past, believed to be dead, are being made to speak, in the course of which it turns out that they propose things altogether different from the familiar, worn-out trivialities they had been presumed to say." No wonder Heidegger's name spread throughout Germany "like the rumor of the hidden king," Arendt wrote, one who "reigned in the realm of thinking."[3]

In 1924, Heidegger was thirty-five years old and not yet a full professor. He had just moved to Marburg after four years as Husserl's assistant. But like many of Husserl's intellectual offspring, Heidegger took phenomenology in a new direction, one that would profoundly influence Arendt. Husserl had assumed that the researcher who bracketed, in Husserl's special sense—or who was familiar with the Husserlian version of the Great Story—could achieve a neutral and objective stance to describe "the thing itself." Heidegger was skeptical and thought bracketing only transformed rather than purified the way a researcher grasped a thing. In his eyes, Husserl was still making pronouncements from "on high," a supposedly neutral scientific position which is not one. Husserl's seemingly neutral descriptions were still beholden to a closet naturalism. If we pay true phenomenological attention, Heidegger thought, things in our world scream out against the bracketed way of encountering things.

In one of his classes, for instance, Heidegger asked students to consider his lectern. What they do *not* see is flat brown surfaces attached at right angles which they interpret as a lectern. No! When they enter the familiar classroom environment the lectern shows itself "in one fell swoop" amid a familiar world of blackboards, books, professors, buildings, and the university. Students belong to *that* world first, which "does not consist just of things, objects," and the lectern emerges from it. Would they see the lectern more fundamentally if they bracketed that world? No! That would distort the phenomenon, making the lectern appear abstractly and artificially. Asking students "*What* do you see?" of just the lectern is thus not a good philosophical question, because it already artificially suppresses the world they inhabit. In recovering the role of the world to which humans first belong, and in which the things they encounter first appear, Arendt wrote, Heidegger "was actually attaining 'the things' that Husserl had proclaimed."[4]

Heidegger knew how to avoid reducing the world to a collection of things with properties, but the same was not true of other Western philosophers. They had a hard time talking about the "world"—the meaning of what Heidegger called "Being"—tending to reduce it to a collection of objects, or a super-object with super-properties that we can come to "know." Western philosophy, Heidegger feared, seemed determined to forget about Being, heading itself into an unprecedented crisis of worldlessness. It is easy to forget, the way speakers of a language tend to forget that "language" is more than words, sentences, and rules but involves a meaning-giving horizon for each act of expression.

Western philosophers, Heidegger thought, kept trying to describe some permanent, unchanging "reality" apart from the world, or consisting of only a small piece of it—to explain all existence. They've tried this trick—the "reality trick," one might call it—from the beginning of Western philosophy, seeking to find Being in some deep reality: God (medieval theologians), a principle of presence grasped clearly and distinctly (Descartes), or the Absolute (Hegel). But Being cannot be turned into a thing. Not only that, Heidegger thought, but each philosopher also implicitly recognized this, a recognition which in his wrestling matches he was forcing them to reveal. He called his procedure "destructive retrieval." It's a retrieval, because he was recovering the sense of Being and world, and destructive, because he has to battle previous philosophers to do so, exposing incoherences in what they say explicitly.

Why had the Western tradition lost the sense of Being and world?

One culprit, Heidegger (like Husserl) thought, was the cultural impact of science itself. The early figures of the scientific revolution had regarded science as motivated by human life concerns, and the products of science as contributing to them. But the success of the sciences was built on a version of the reality trick, trying to explain the world in terms of a small number of elements. The successes of this trick fostered the idea that the world is a collection of independent things, and that knowledge is a collection of information about them. The reality trick therefore came to strongly shape Western views of reality itself.

Heidegger's procedure of destructive retrieval countered the reality trick, focusing the attention of philosophers on experienced phenomena. This had an extraordinary impact on his students. The philosophical tradition, he told them, was not bygone; it had ushered us into the present in a way that opens some doors and closes others. Heidegger did not want his students to let the concepts they had inherited rule their thinking; he wanted them to use experience—always the phenomenological first principle—as their guide to new experiences. Lived experience is the source of the urge to be creative, but when it is objectified it produces a narrow and anemic account of phenomena. Heidegger's students had to take over their world as the tradition had delivered it to them—their "thrownness," as he called it—to think by exploring the difference between what they understood implicitly and what the tradition told them it was acceptable to say, and then get the tradition to help them say something new. In this process they would never arrive at some neutral, final ground (if they did, they would have become traditional metaphysicians), and their questioning would be at the same time self-questioning. Students felt humbled and liberated. As Arendt would put it, quoting Heidegger himself, "We left the arrogance of all Absolute behind us." The power that swept through Heidegger's thinking, she concluded, came from "the primeval." It was a "*passionate* thinking, in which thinking and aliveness become one."[5]

Arendt's first course with Heidegger was in the summer semester of 1924. Among her classmates were many who would soon become philosophical luminaries themselves, such as Hans-Georg Gadamer, Hans Jonas, Herbert Marcuse, and Leo Strauss. Jonas, who like Arendt was Jewish, fled Germany in 1933 and ended up teaching at the New School for Social Research in New York. He once recalled the impact Arendt herself made on this illustrious group:

> Shy and withdrawn, with strikingly beautiful features and lonely eyes, she stood out immediately as exceptional, as unique in an as yet indefinable way. Brightness of intellect was no rare article there. But here was an intensity, an inner direction, an instinct for quality, a groping for essence, a probing for depth, which cast a magic about her. One sensed an absolute determination to be herself, with the toughness to carry it through in the face of great vulnerability.[6]

For Arendt, it was Heidegger who was magical. Her biographer Elisabeth Young-Bruehl writes,

> When Hannah Arendt encountered Martin Heidegger everything changed. He was a figure out of a romance—gifted to the point of genius, poetic, aloof from both professional thinkers and adulatory students, severely handsome, simply dressed in peasant clothes, an avid skier who enjoyed giving skiing lessons.[7]

Her love for Heidegger "abruptly and frighteningly, ended her youth, her innocence."[8]

Arendt and Heidegger began an affair in 1925. He was thirty-five, of Catholic upbringing, and married with two children; she was eighteen,

Jewish, and a single, first-year university student in love for the first time. They kept the affair secret even from their closest friends, meeting in her attic room not far from the university. In the fall of 1925, she spent a semester in Freiburg studying with Husserl. Knowing that her personal relationship with Heidegger made it impossible for him to supervise her dissertation, she moved to Heidelberg to study with Heidegger's friend Karl Jaspers. For several years she continued to meet Heidegger for trysts.

MORAL EXAMPLE

If Arendt found Heidegger a model of how to think, she found Jaspers a model of how to be a human being. Though an "existential" thinker, Jaspers preferred to address contemporary issues rather than the history of philosophy. As Young-Bruehl notes, Jaspers was everything Heidegger was not: "a moral example, a cosmopolitan, a model of the public philosopher."[9] As someone who grew up fatherless, Arendt told an interviewer, she was enormously influenced by Jaspers as a mentor; "if anyone has succeeded in instilling some sense in me, it was he."[10] Her future work would combine the historical depth and phenomenological orientation of Heidegger's approach and the sensitivity to and necessity of confronting contemporary issues that Jaspers embodied. Jaspers had been a regular at Weber's Sunday afternoon salons, which Marianne continued after Max's death in 1920. Jaspers was inspired by Weber's insightful method of using ideal types and worldviews to understand the murky flowing river of human affairs. At the same time, Jaspers shared Heidegger's more experiential concern and thought that the way humans lived through that flow had to do with why they took on the specific ways of existing they did. A true philosophy infused the way a person acted and lived. Partly for this reason, Jaspers was dissatisfied with Heidegger's book *Being and Time*, which appeared in 1927; it seemed remote from the actual issues of the age. His relationship with Heidegger grew more strained when he heard reports of Heidegger's support for National Socialism.

Arendt spent four years in Heidelberg writing a dissertation under Jaspers on St. Augustine's several conceptions of love. The dissertation, a Heideggerian-inspired wrestling match in which a Jewish woman practiced destructive retrieval on a revered Christian saint over a sacred issue, caused a stir in Heidelberg. Young-Bruehl notes, "Arendt began her publishing career as she ended it more than forty years later—as a burr under scholarly saddles."[11]

While writing the dissertation, Arendt expanded her circle of friends and concerns, and became increasingly interested in "the Jewish Question," or the matter of how Jews were treated in different countries. While not a Zionist herself, she became a close friend of Kurt Blumenfeld, the president of the Zionist Federation of Germany. At a masked ball in Heidelberg, dressed as an Arab harem girl, she met Günther Stern, another student whom she had last seen in Heidegger's 1925 Marburg class. The two moved into a one-room apartment they had to share with a dancing studio. They married in 1929.

In 1930, the couple moved to Berlin, where Günther became a journalist, worked on a novel, and unsuccessfully pursued a *Habilitation*, a step toward a university professorship. He changed his last name to Anders, or "Other," to sound less Jewish for his journalistic ambitions. Arendt wrote essays and reviews, including one for Weber's journal the *Archiv*. For her *Habilitation*, she began work on a biography of Rahel Varnhagen, an assimilated Jewish woman who ran a salon a hundred years earlier, and to whom Arendt felt a special kinship. Varnhagen had loved and lost, felt homeless, and was forced to live in an anti-Semitic world. She, too, had known no comfortable and solid cultural ground. Arendt referred to Varnhagen as her "closest friend, though she has been dead for some hundred years."

Arendt's biography of Varnhagen was a self-discovery. Arendt began to realize the costs of assimilation; in Europe, as she put it, "it is possible to assimilate only by assimilating to anti-Semitism also."[12] She saw that Germany's treatment of Jews was not a quirk of German history but a pan-European phenomenon, for how a country treated Jews illuminated "the ugly reality of the gaps in [its] social structure."[13] She realized that being Jewish was not just a feature of one's private life but a public fact to which personal concerns—culture, language, beliefs, worship—were irrelevant. This public fact made even assimilated and converted Jews vulnerable to abuse, arrest, and annihilation. When attacked in public, then, it was useless to appeal to theories or ideologies or principles; one had to respond—to act—by assuming that public identity. "If one is attacked as a Jew, one must defend oneself as a Jew," she told an interviewer. "Not as a German, not as a world-citizen, not as an upholder of the Rights of Man, or whatever. But: What can I specifically do as a Jew?"[14] This alerted Arendt to the distinction between the private and public spheres, the latter a place in which citizens can appear as they are to others and freely act. Heidegger insisted each person had to embrace one's thrownness, the world delivered them, if one were to truly think. Arendt now realized that one had to embrace one's thrownness, in the sense of one's public identity, if one were to truly act. Accepting her Jewishness was not an obstacle to her genuinely acting in the world, as per Kant, but rather its condition.

Arendt began to work with Jewish organizations and joined a Zionist organization. Not because she was a Zionist—she wasn't—but because "they were the only ones who were ready."[15] She spent more time with her Zionist friend Blumenfeld, another independent spirit, who introduced her to Cuban cigars, which she smoked in public over her husband's protests. But the growing momentum of National Socialism made life more difficult. "We were very poor, we were hunted down, we had to flee, by hook or by crook we somehow had to get through."[16]

Arendt dated her "turn to the political" to February 27, 1933. That day an arson attack on Germany's Parliament building, the Reichstag, set in motion events that completed the Nazi ascent to power. Though Hitler was chancellor, the Nazi Party did not yet have a majority in Parliament. Hitler now blamed the fire on Communists and used it to justify mass arrests that included Communist members of Parliament, making the Nazis the majority. The ensuing political repression was "monstrous," Arendt said later, "an immediate shock for me." It represented the drying up of what she would soon call public space, in which genuine human action is possible.

Another shock was how willing German intellectuals were to work with Nazis. Arendt called this *Gleichschaltung*. It is usually translated as "coordination," but "normalization" is better. The impulse to normalize, to sit down and work with Nazis, was rapidly making it harder for their opponents to act. This made her fear, for the rest of her life, the ability of intellectuals to compromise their thinking and let themselves be caught up in political currents. It solidified her determination to be uncompromising, a burr under saddles if necessary.

Günther fled to Paris a few days after the Nazi takeover, afraid his name would turn up on some list as an opponent of the regime. Hannah stayed. "I was no longer of the opinion," she remarked, "that one can simply be a bystander." Blumenfeld asked her to secretly collect evidence of anti-Semitic statements to use to convince the outside world of the rising danger. It was incredibly risky; if caught she would surely be charged with acting against the state. On her way to meet her mother at a Berlin café one day, she was caught and jailed, but managed to charm the relatively inexperienced arresting officer into being released after eight days.

Knowing that she might not continue to have such luck, Hannah and her mother fled their native Germany and walked across the border to Czechoslovakia, where they took a train to Geneva and then Paris. Hannah reunited with Günther, bringing the manuscript of his unfinished novel which he had left behind in Berlin, still smelling of the smoked bacon in which it had been hidden for months. "At least I am not 'innocent,' " she told an interviewer later, proud of her illegal actions against the Nazis. "No one could say that of me!"[17]

STATELESS

Arendt was now twenty-seven years old and stateless, her condition for the next eighteen years. She found Paris clogged with wave after wave of refugees, all seeking housing and jobs and facing hostility from the thousands of unemployed Parisians. She and Günther scraped by with difficulty but grew apart. He soon left for New York, and in 1937 the two amicably divorced.

In Paris, she joined a network of friends who held what amounted to an ongoing seminar, generally in the apartment of Günther's cousin, the German philosopher Walter Benjamin. There Arendt met her future husband Heinrich Blücher, a former Communist who had also fled from Berlin to Paris. This growing "tribe," as she called it, helped sustain its members in the absence of a public space; Young-Bruehl writes that in it "there was no place for the national differences, cultural barriers, ideological clashes, or class conflicts that gave the European world around it, as Arendt said, 'the sordid and weird atmosphere of a Strindbergian family quarrel.' "[18]

Arendt began working for a refugee organization that helped Jewish teenagers flee Germany for Palestine, providing them with clothing, food, education, money, and papers. In 1939, Blücher was sent as an alien national to an internment camp. A few months later he was freed; but in 1940, after Germany started to invade Belgium, both Hannah and Blücher were sent to separate internment camps. Each managed to escape. They found each other and obtained visas to America. They were luckier than many friends; Benjamin committed suicide after being apprehended at the Spanish border and fearing he would be returned to

Nazi-occupied France. In April, Arendt and Blücher left from Lisbon on a boat to New York.

NEW YORK

The couple arrived in New York with $25 in their pockets and a promise of a $75 monthly stipend from a Jewish organization. Arendt rented two small apartments on the Upper West Side, one for her and Blücher, the other for her mother, due to arrive shortly. She supported herself with administrative jobs and wrote articles. Her knowledge of English was spotty and for a long time she used a translator. But the Jewish community was organized, its venues were respected, and prestigious journals wanted to publish her articles, an opportunity she hadn't had in Europe. In America, she wrote, "assimilation is not the price of citizenship."[19]

Blücher was unemployed, and Hannah supported him and her mother, who kept nagging that her husband should do more for the family. They had no children; Hannah told a friend later, "when we were young enough to have children, we had no money, and when we had money, we were old."[20] They began to collect a new "tribe," whose members widened after Hannah got a job as a senior editor at Schocken Books, a German publishing house that had been closed by the Nazis but reopened in the United States. One tribal chieftain whom Hannah met through Schocken was the novelist Randall Jarrell, whose book *Pictures from an Institution* loosely modeled the cantankerous couple "the Rosenbaums" on Arendt and Blücher. You "never knew what the Rosenbaums would talk about, or what they would say about it when they did," the narrator says, describing Irene Rosenbaum as "a slight, animated, disquieting woman" with "frighteningly deep eye-sockets" who "spoke her mind, shortly and firmly, about anything and everything."

Arendt started writing articles shortly after she arrived in New York;

she wrote to understand, she'd say. One of her most moving essays was "We Refugees," for a small journal of Jewish culture called the *Menorah Journal*. It's devastating for its simplicity, insight, and directness. "In the first place, we don't like to be called 'refugees,' " Arendt begins. That term implies we committed some dreadful act or held some radical idea; in fact we did nothing of the kind; most of us are as far from radical as you could get. Yet,

> we lost our home, which means the familiarity of daily life. We lost our occupation, which means the confidence that we are of some use in this world. We lost our language, which means the naturalness of reactions, the simplicity of gestures, the unaffected expression of feelings. We left our relatives in the Polish ghettos and our best friends have been killed in concentration camps, and that means the rupture of our private lives.[21]

When you are a refugee, friends commit suicide without explanation; none is needed. People shed their identities and adopt new ones "so frequently that nobody can find out who we actually are." Arendt's essay provides a moving description of statelessness—or what might be called political worldlessness—in which people are stripped of traditions and connections.

In another essay written toward the end of the war, Arendt wrote of how devastated the public space had become under Nazism. While a principled country seeks to make legal and moral behavior compatible, in Nazi Germany legal behavior—behavior that was unpunished by the state—required becoming an accomplice to the "vast machine of administrative mass murder." Politicians displayed an obsession with loyalty, opportunism and private interests flourished, and true patriotism all but evaporated, replaced with empty words about the greatness of the Fatherland.[22] As a result, the "classic virtues of civic behavior" vanished, leaving no room for what Aristotle called practically wise action informed by the best available assessment of the potential for action in the world.

ORIGINS OF TOTALITARIANISM

"Statelessness," writes Young-Bruehl, "taught her the elements of political life."[23] Shortly after "We Refugees," Arendt embarked on what would turn into her first book, *The Origins of Totalitarianism*. It was not a history of events or ideas, but an exploration of the roots of what she had just lived through, the emergence of totalitarian regimes in Nazi Germany and Stalin's Russia. The structure of these seemingly vastly different regimes was strikingly similar. They each deliberately enforced worldlessness on millions of their own inhabitants by using the same tools (camps, genocide) and by exploiting racism and anti-Semitism. Nothing on this scale had ever happened before, yet Arendt felt the development could be explained after a destructive retrieval-like exploration of Western thinking about politics. While Heidegger was trying to understand the space in which authentic thinking could occur, Arendt was trying to understand the public and political space of authentic action. Arendt's intention was not to describe the human toll of totalitarianism, nor its failures as a political theory, nor yet to denounce or moralize about it—which would be ineffective—but to understand how it came into being in the first place.

The Origins of Totalitarianism was published in 1951 and had three parts: Antisemitism, Imperialism, and Totalitarianism. The first is sometimes called the best brief account of European anti-Semitism; Arendt treated it not as a torment particular to European Jews but put it at what one historian called "the storm center of events."[24] Furthermore, she saw modern anti-Semitism not as a scaled-up version of what had come before, but as qualitatively different.

The second part, Imperialism, examined the development of the ideas of race and of rights. Until the end of the eighteenth century, what rights humans had were regarded as grounded in religious or social traditions, or existing by government decree ("bills of rights"). Then came a "turning point in history," the appearance of "The Declaration of the Rights of Man" during the French Revolution (1789). That document portrayed human rights as universal and inalienable, deriving their authority not from governments or humans but nature itself. The idea that humans had natural rights was old; what was new was using this idea as a tool against governments who justified themselves by appeals to nature. Yet the plight of stateless people showed natural authority to be all but worthless; humans effectively only had rights in a public space in which institutions guaranteed such rights. "Not the loss of specific rights, then, but the loss of a community willing and able to guarantee any rights whatsoever, has been the calamity which has befallen ever-increasing numbers of people."[25] Arendt cleverly called the right to have access to such a public space the "right to have rights," and called what humans gain when they have such a space not freedom or justice—terms she found too abstract—but the capacity for action. She defined action as the ability to begin something genuinely new.

The third part, Totalitarianism, has been described as "a phenomenology of totalitarian domination." It outlines what happens when the collapse of traditional political authority described in the second part of her book couples with the anti-Semitism and racism described in the first, carried out with the thoroughness of Weberian rationalization. Millions of people end up trapped in a political version of the iron cage, treated as "not belonging to the world at all, which is among the most radical and desperate experiences of man." Two lies are offered to support totalitarianism. One is that political societies are products of human engineering just like houses and boats. The second is that totalitarianism is an inevitable consequence of historical processes—for Nazism, the triumph of the Aryan race; for Marxism, the triumph of the classless society. Arendt called totalitarianism a "radical evil" that makes human beings "superfluous," disposable pieces in a system where concentration camps "are the laboratories where changes in human nature are tested" and where "total domination" of humanity is achieved.

Arendt's account of totalitarianism gives Vico's description of barbarism a twist. Vico had taken the public space of the human world, and the ability to act, for granted. From the early poetic stage to the culminating rational one, Vico's humans always had space in which they acted and could transform from the inside, driving the historical process. In the rational stage, humans are indeed prone to greed, egotism, and acting like businessmen to the point where they begin to act like barbarians; still, they remain political actors. Arendt's vision was more dire and extreme: the historical process has made it possible for nations to eliminate the public space itself, leaving humans no room to exercise practically wise action. But this possibility is not inevitable. Exposing the elements that helped bring it about can make it possible to avoid in the future. More concretely, you cannot prevent another Holocaust or gulag by moralizing, but only by understanding and exposing what had produced the totalitarianism that gave rise to these things. Much of Arendt's writings explores this public space, its conditions, its relation to other spaces, and the consequences of its loss.

Here and elsewhere, Arendt practiced a different kind of storytelling. Jewish history, she wrote, was made not by Jews but by people around them, and "appears at first glance to be a monotonous chronicle of persecution and misfortune, of the brilliant rise and fall of a few individuals, atoned for by pogroms and expulsion of the masses."[26] Instead, the key forces in the story are not just anti-Semitism but the collapse of authority, the rise of rationalization, and the immense power of bureaucracy thanks to modern science and technology.

CELEBRITY

The Origins of Totalitarianism was a thoroughly documented account that explained to Americans the background of what was not yet widely called the Holocaust. Overnight it made Arendt a public intellectual, and among other things landed her on the cover of the prestigious *Saturday Review*.[27] Thanks to her new celebrity, and Blücher's getting a teaching job at Bard College, the couple was able to move to a larger apartment.

In December 1951, Arendt passed the exam to become an American citizen. To prep, she had "a self-taught course in the political philosophy of the Founding Fathers." She appreciated that the Fathers intended not only to guarantee American citizens—those not women and slaves at least—access to a shared public space, but to provide them with tools to support acting in it. She was fascinated, for instance, by Jefferson's idea of a "ward system," or networks of citizens who cooperate locally to encourage people to overcome private interests and act in concert in the public space. While America's founders saw themselves as emulating Rome's founders, Arendt thought, their real achievement—their greatness—lay not so much in creating new structures as in securing a participatory space in which American citizens could act.

In the next few years, Arendt explored the logic of totalitarianism in numerous essays, usually in response to specific events. "The Eggs Speak Up" wittily uses the proverb that "You can't make an omelet without breaking eggs" to expose the cynical justification used to defend repression in the Soviet Union; the eggs are humans and the omelet the state. Yet the essay also scorned those—ex-Communists in particular—who unthinkingly and with a superior attitude championed the democratic over the Soviet system. "Democratic society as a living reality is threatened at the very moment that democracy becomes a 'cause,' because then actions are likely to be judged and opinions evaluated in terms of ultimate ends and not on their inherent merits." What we should extol as great is not democracy but the space for action that its institutions grant, action meaning not just what brings about desirable outcomes but unpredictable creative forays.

Arendt elaborated these thoughts in a scathing review of the memoir *Witness* (1952) by American writer and Soviet spy Whittaker Chambers. In the 1930s, Chambers enthusiastically embraced Communism as the wave of the future, but he then changed his mind and equally enthusiastically embraced anticommunism. Politics is messy and full of uncertainty, Arendt wrote; this is one reason why philosophers tend to prefer the contemplative life in which one studies eternal things. We should therefore fear actors in the political realm like Chambers who are sure that they Know the Course of History and want to take the rest of us along. "It is against these makers of history that a free society has to defend itself," she wrote, adding, in a sentence with contemporary resonance, "If you try to 'make America more American' or a model of democracy according to any preconceived idea, you can only destroy it." To do so, she concluded, would only "strengthen those dangerous elements which are present in all free societies today and which we do not want to crystallize into a totalitarian movement or a totalitarian form of domination, no matter what its cause and ideological content."[28]

Publishing such thoughts—the essay was reprinted by the *Washington Post*—was career-threatening at a time when Senator Joseph McCarthy was leading a witch hunt against suspected Communists and Communist sympathizers, and with Arendt's own husband a former Communist (a fact he had concealed from immigration officials). Yet it was also a sterling example of action in the very sense that Arendt wanted to develop: responding to the world as she encountered it in a way that, in concert with the actions of others, could protect and rejuvenate the public space.

LOVE OF THE WORLD

In the 1950s, Arendt began to follow up *The Origins of Totalitarianism* with an extended critique of Marxism. It slowly turned into the most systematic treatment of her ideas about the public space and the dangers threatening it. At first she wanted to call it *Love of the World*, but its final title was *The Human Condition*.

The human condition, she wrote, is shaped by three capacities. One is labor, such as in factories and fields, which brings humans the bare necessities of life. Another is work, or making and maintaining the institutions that shape the human world. A third is action, the ability to initiate genuinely new things in concert with others, and to change that human world. Action, rather than needs or interests, is what politics and democracy are ultimately about. But action is only possible in a public space where one can initiate new processes in concert with others.

Action is endangered in the modern world. Totalitarianism is but the most dramatic and ugly example of its near extinction. In democracies action is threatened by the social sphere, in which public life is ruled, not by the desire to act in concert with others, but by fashions and self-interests as if the nation were a giant household or business. The triumph of the social sphere involves the "degradation of all things into means, their loss of intrinsic and independent value."[29] Europeans call this process "Americanization," but Arendt saw it as really modernity.

Philosophers have not helped to understand these developments. They tended to look down their noses at politics and action, as Arendt knew only too painfully from her experiences with Heidegger and other Nazi normalizers. In *The Human Condition*, Arendt traced the disdain all the way back to Plato's cave image at the origin of the Western philosophical tradition.

Plato's cave—metaphorically the *polis* or public realm—is a dim, unruly place where humans are enchanted by shadows and misled by illusions cast by their leaders. True leadership, for Plato, would require imposing the eternal truths found outside the cave. But a philosopher who has been outside to see them would be blinded and confused on entering the cave, and mocked and feared by its inhabitants, who might even threaten the philosopher's life (Plato's reference to Socrates's trial and execution). No wonder philosophers prefer to remain outside, contemplating the eternal. Heidegger had seen in this image the beginning of the Western obsession with the eternal and unchanging, with its suspicion of change and becoming, and with the theory-happy behavior of applying rules and principles to life rather than allowing them to emerge through experience. Arendt now saw in this image the beginning of the disdain for politics and action as well. As the scholar Richard H. King put it, "If Martin Heidegger located the forgetfulness of Being at the very source of Western philosophy, Arendt identified a forgetfulness of the world at the origins of Western political thought."[30]

For Arendt, the cave image was the birth of "political philosophy," the idea (in her mind, mistaken) that politics was a matter of trying to clean up the unruly public space by finding and using the right theories and following the rules. This made political life an engineering project, with theories—including not only Marxism but all other ideologies—providing construction manuals. It is, she says, the escape from action into rule.[31] But such theories are drawn from a standpoint above life—from an audience or spectator's position—not from onstage, from within life. What Arendt thought of as genuine political life, or what she called the *vita active* or "active life," was an attuned embracing of our identity in the world that consisted both of acting in the polis and helping to preserve it. "The work of politics becomes twofold," writes the philosopher Anne O'Byrne, "consisting, on the one hand, of acting, speaking, and appearing before one's peers and, on the other, of remembering, telling stories, and preserving laws so that the *polis* itself might be preserved."[32]

In the final section of *The Human Condition*, "The *Vita Activa* and the Modern Age," Arendt saw the decline of the public sphere, which began with Plato and continued with a hidden Platonism disguised in such imagery as "applied science" and "social engineering," as coupling in a disturbing way with transformations in science and technology. These developments began with Galileo's telescope, which had an extraordinary impact not because it confirmed heliocentrism by itself, but because now "the secrets of the universe were delivered to human cognition 'with the certainty of sense-perception.' "[33] Earth-bound humans with their body-tethered senses could now access secrets of nature as confidently as any other truths. This provided the template for what Arendt calls "natural science," which sought to discover the forces of nature and ways to control them. Natural science began to put an ever greater distance between humans and the Earth, "alienating man from his immediate earthly surroundings,"[34] and encouraging humans to regard themselves as the Earth's "lord and master," as Descartes had put it in the *Discourse*. Its legacy, that is, involves two related assumptions: that to be rational animals is to be mathematizing thinkers able to control nature (masters), and that such control leaves us free to control it for our comfort and satisfaction (lords). The practice of science came to be regarded, not so much as intervening in nature (in Bacon's terms) to get it to reveal itself in preparation for harnessing its forces, but as taking a distanced, audience's perspective on it, or what Arendt calls an Archimedean point, adopting "a universal, astrophysical viewpoint, a cosmic standpoint outside nature itself."[35] This is the science that Descartes dreamed of centuries before in his view from the boat; now, Arendt says, it is the perspective of all science.

Arendt saw this perspectival shift as culminating in "world alienation," whose symbol was Sputnik, the first Earth satellite launched a few months before Arendt completed *The Human Condition*. The irony is that this Archimedean point outside the Earth, and those to follow, were constructed by Earth-bound creatures.[36] It is a perspective whose symbolic stature is now shared by such things as genetic engineering and nanotechnology. The new power of science and technology that accompanies this shift is vastly different from the old. Instead of merely harnessing natural forces, universal science now "acts into" nature, initiating new processes such as nuclear energy and genetic modifications. Like all action, this is only achieved in concert with other humans, and introduces uncertainty into the world—the possibility of Frankensteins.

Husserl regarded the cultural crisis around him as due to an antirationalism provoked by the forgetting of the origin of the sciences. Arendt saw something else sinister: loss of the texture of reality itself, which harbors the possibility of "the deadliest, most sterile passivity history has ever known."[37] She would surely have regarded the control that a small number of individuals now exert over media, finance, economics, technological developments, and political life—the form that totalitarianism can take in democracy—as another stride toward that passivity. This loss of texture allows incompetents like Eichmann to dominate what's left of the public space, setting themselves up as umpires not simply of what's allowable, but also of what's just and what's scientific.

AUTHORITY

Authority is a key feature of public space, Arendt thought, and she saw it, too, as waning. The success of totalitarianism is only possible, Arendt thought, with the erosion of authority. Totalitarianism involves coercion and force, not voluntary deferral; "authority implies an obedience in which men retain their freedom."[38] But authority also disappeared in the social space of democracies. Arendt found that "the disappearance of practically all traditionally established authorities has been one of the most spectacular characteristics of the modern world."[39] She explored the consequences in several essays.

One was in education. In "The Crisis in Education" (1958), Arendt noted the difference between a teacher's qualifications and authority. A teacher's qualifications involve knowing something and how to teach it, while authority involves assuming responsibility. That's similar to the way the authority of a mountain guide comes not just from knowledge of terrain but also an assumption of responsibility for those guided. The teacher-student form of authority, along with that between adults and children, which Arendt discussed in another essay called "What Was Authority?," has long provided the model of political authority. Now that such authority is distrusted, it threatens the ability to pass on culture, but also deprives humans of "depth in human existence." She then embarked on another destructive retrieval, trying to understand how the idea of authority emerged. This, too, goes all the way back to Plato, who pondered how to guide humans without resorting to coercion on the one hand or mere persuasion on the other. Reason works, but for the few rather than for the many. Plato's problem, which he did not entirely solve, was to find a relationship between ruler and ruled "in which the compelling element lies in the relationship itself and is prior to the actual issuance of commands," the way a patient accepts the doctor's authority because of their relationship—which is motivated ultimately by the concern for one's own health—rather than coaxing or scientific proof.[40]

Weber had pointed out that authority needed to be sought. Arendt now wondered what the compelling element might be that would motivate people to seek political authority. She did not entirely succeed. We have to accept, she wrote, that it could no longer come from a sacred source, a transcendent origin, or traditional standards of behavior. But this then requires us to reopen "the elementary problems of human living-together."[41] The closest she came to identifying that compelling element was love of the world itself.

The episode I mentioned at the beginning of the chapter on Descartes, about the scientist who explained that he loved the numbers because he loved his grandson, is a good illustration of the dynamics of science authority and denial as they might be reconstructed from Arendt's work. The scientist's love was the compelling element that bound him to the numbers and made them authoritative. The activists lacked that compelling element. They regarded those who produced the numbers as responding not to the world as a whole but to some piece of it in the workshop environment. Theirs was the general suspicion of scientific authority often encountered in debates over the safety of nuclear power, genetically modified organisms, vaccinations, and other public controversies with a scientific-technological dimension. The episode also illustrates what would have to happen to counter that suspicion, namely, a recognition on the part of nonscientists that those who produce numbers in the workshop are not only assuming responsibility, but also that the numbers are of vital concern for the nonscientists' own love of the world. The task in countering science denial is discovering how to promote that sense in an age of world alienation. That would require fostering an environment in which the numbers mattered—the "right to have science," one might say.

What the crumbling of authority has done to scientific authority in the public realm can be compared to what Arendt said about totalitarianism's impact on the public space. Just as a genuine political space seeks to prevent a disparity between legal and moral activity, so it seeks to allow practically wise action to be scientifically informed—to have the "right to have science." The danger of totalitarianism is that it eliminates the public space, leaving humans no room to exercise practically wise action; the danger of a political atmosphere that permits science denial is to leave no room for scientifically informed, practically wise action. Science then lacks authority in the public realm and becomes just another political tool; those who appeal to science to justify their actions are then inevitably accused of being politically motivated. All authority becomes the authority of motives. The political atmosphere that permits science denial is then allowed to persist by the inevitable process of *Gleichschaltung* or normalization.

Just as it is hopeless to counter totalitarianism by denouncing or exposing individual instances of guilt, because what makes the gap between the legal and the moral possible is the political atmosphere, so it is hopeless to counter science denial by denouncing or exposing individual instances of it. This condition then delivers the ability to make climate-threatening decisions for unprepared people.

What if science is genuinely in the service of special interests, or is too technical for politicians, or the jury is still out? Here, too, it is fruitful to compare totalitarianism and a state that tolerates science denial. Just as a genuinely just state seeks to correct injustices and to rectify disparities between the legal and the moral rather than allow them to persist, so the genuinely wise state seeks to correct misuses of science, convert abstract information into information that is practically useful, and deliberate about scenarios, allowing practically wise action to be scientifically informed.

DEFACTUALIZING

Arendt published her thoughts on authority in *Beyond Past and Future* in 1961. Her next book was *Eichmann in Jerusalem* (1963), based on articles for the *New Yorker* covering the trial of the Nazi war criminal Adolf Eichmann for crimes against humanity. While she concluded that Eichmann "must hang," her judgment that he—an insatiable self-promoter and braggart who coveted media coverage—was an emblem of modern totalitarianism, as ordinary as the average German who had thoughtlessly followed orders, was highly controversial. Her views, though, had sprung out of the ideas she had been developing all along.[42] She tried to shift the focus from the question of Eichmann's individual guilt (and even that of the Nazis), and to reconsider responsibility and the Holocaust productively to enable her to question, non-moralistically, how to prevent future Holocausts and gulags.

But her protective shell was sorely battered by critics of *Eichmann in Jerusalem*, who sometimes distorted facts in their campaign against her book. She wrote an essay called "Truth and Politics" largely in response, though it nowhere mentions Eichmann. The essay is uncannily relevant to science denial. Truth and politics, she began, "are on rather bad terms," with an old and complex relationship that is uncomfortable to inquire into. Never before have so many different religious or philosophical opinions been so tolerated, so it might seem that truth could find a niche somewhere in the public sphere. But this is only the illusion of toleration; these religious beliefs and philosophical opinions have become empty, not grounded in experiences that make such beliefs and opinions worth having. The confirmation is that "factual truth, if it happens to oppose a given group's profit or pleasure, is greeted today with greater hostility than ever before."[43] Also disturbing is the tendency to counter uncomfortable facts by turning them into opinions, and then opposing these opinions to others.

But it is false to put facts and opinions on a par in this way. Factual truth, Arendt says, is already social; "it concerns events and circumstances in which many are involved; it is established by witnesses and depends upon testimony; it exists only to the extent that it is spoken about, even if it occurs in the domain of privacy." Yet this very social character also makes it vulnerable to being discounted. Factual truth is essential to the public space and the ability to act. "Freedom of opinion is a farce unless factual information is guaranteed and the facts themselves are not in dispute." She concludes: "Conceptually, we may call truth what we cannot change; metaphorically, it is the ground on which we stand and the sky that stretches above us." To threaten facts is to threaten human existence, politics, and freedom itself.

A few years later, a dramatic event—a key episode in the relationship between the US press and the government—prompted Arendt to return to the topic of lying and deception. In June 1971, US newspapers began printing excerpts from the "Pentagon Papers," a purloined copy of a forty-seven-volume study of the US participation in Indochina, defying the government's unsuccessful attempts to block publication. That summer, Arendt, still mourning Blücher's death the year before, spent a month at the vacation home of her old friend Mary McCarthy in Castine, Maine, living in an apartment over the garage. There she wrote "Lying in Politics: Reflections on the Pentagon Papers," which the *New York Review of Books* printed that fall. This essay has a still uncanny relevance to science denial.

The Pentagon Papers, she wrote—written by the government for government eyes only—provide a valuable insight into government deception and lying. "The relation, or, rather, nonrelation, between facts and decision, between the intelligence community and the civilian and military services, is perhaps the most momentous, and certainly the best-guarded, secret that the Pentagon Papers revealed." The document indicated that while the intelligence community had produced "astoundingly accurate factual reports," the Washington bureaucrats had lied about them. No surprise here: "lies have always been regarded as justifiable tools in political dealings," for they can preserve possibilities for action and the political space. "The lie did not creep into politics by some accident of human sinfulness."[44] Furthermore:

> Lies are often much more plausible, more appealing to reason, than reality, since the liar has the great advantage of knowing beforehand what the audience wishes or expects to hear. He has prepared his story for public consumption with a careful eye to making it credible, whereas reality has the disconcerting habit of confronting us with the unexpected, for which we were not prepared.[45]

Political lies are effective, that is, because they are not just untruths but bespoke untruths, targeted and adapted for specific audiences to accomplish specific ends. Facts, on the other hand, seem to come out of nowhere and possess an "infuriating stubbornness."

But the Washington bureaucrats had lied in a new way. Their goal was not to contain some enemy but to protect their image, aiming at no more than getting their audience to buy in to their agenda. They saw themselves as effective problem solvers who resolved problems by applying military theories. To do so they had to "defactualize," obliterating facts that stood in the way of those theories rather than consulting experience and learning from reality. "The problem solvers did not judge; they calculated."[46] In the absence of the relevant facts their calculations resulted in disaster.

With just a few terminological switches, Arendt's essay offers startling insight into the dynamics of science denial. For intelligence community read scientific infrastructure. For containing the enemy read coping with nature. For protecting one's image read defending one's ideology. For military theories read economic notions. For Washington bureaucrats read Washington bureaucrats. The core problem is "the inability or unwillingness to consult experience and to learn from reality."[47] The great shame is how unnecessary it all is: the principal agent involved in both the Pentagon Papers episode and contemporary science denial is not a banana republic or country desperately seeking to stem a decline, but the most powerful nation on Earth, equipped with the greatest and most reliable of support infrastructures, regarding both intelligence and science.

Arendt's 1971 essay briskly sums up her conception of political space and what is happening to it today. At the end of that year, she discovered she had angina, but this did not stop her from smoking, lecturing, or consuming huge amounts of caffeine. In 1973, she gave the first part of the Gifford Lectures, a prestigious lecture series requiring her to deliver talks at several Scottish universities. She presented the first of three parts of a book she was working on to be called *The Life of the Mind*. In 1974, in the middle of the second part of her lecture series, she had a heart attack and was rushed to the intensive care unit of a hospital.

The day after Thanksgiving 1975, Arendt fell while emerging from a taxi to enter her Riverside Drive apartment; she paused a few minutes before getting up and walking inside. She insisted she was fine and did not see a doctor. Evidently she suffered more than she let on. A week later, in company of friends, she coughed, lost consciousness, and died of a heart attack.

ARENDT AND SCIENCE DENIAL

Arendt's exploration of the dynamics of politics, facts, and truth in the public sphere provides key insights into the authority of science, and science denial.

First, Arendt's writings describe the importance of this issue. Facts are the horizon for effective human activity, the "ground on which we stand." The First Amendment right of free expression is worthless unless such a ground—making possible the ability to state, disseminate, and defend facts—exists. Scientifically established facts about nature are especially important as helping to outline the best available assessment of the potential for action. But facts about nature are not self-authenticating; they are produced in a collective effort and are authoritative only in a space in which human beings already love and respect nature; that's the "compelling element" that underlies the authority of scientists. In such a space, scientific findings can have something like the authority of guides to an unfamiliar environment, but in its absence that authority is lost.

Arendt's description of the authority of the Rights of Man strikingly resembles that of the authority of science. Each is regarded as an indication of the maturation of humanity. Each is regarded as an abstract authority deriving not from human traditions and accidents of history, but rather from nature itself. Each vanishes without a public space protected by human institutions in which that authority is allowed to exercise itself, and a love of the world in which such authority is sought. Just as an equitable society has established a "right to have rights," so a scientifically literate one has established a "right to have scientific authority." When such a space vanishes, ideologies and instincts rule. Lack of such a space cuts human beings off from the ability to make wise political decisions. It fosters wild and pointless behavior, such as Senator Inhofe's attempt to refute what scientists were saying about global warming by bringing a snowball onto the floor of the US Congress, or "Bermdoggle," in which Louisiana governor Bobby Jindal invested $220 million to build sand berms in an attempt to block an oil spill—and when scientists denounced the scheme as ineffective, he compared himself to Huey Long, a populist predecessor, in his defiance of them.[48]

Second, Arendt's writings help clarify what makes scientific findings vulnerable to denial. She discusses the three reasons already mentioned in previous chapters—that scientific facts are generated by bureaucratic institutions, that science is abstract, and that it is open-ended—and adds another reason, fear generated by the fact that science acts into nature.

Bureaucracies. Arendt's remark that scientists do their work in institutions that act and acquire power might make one suspicious that these institutions do not act purely but in the service of their own needs and interests. Scientists have a "political agenda," Donald Trump has said.[49] This is the "It's a conspiracy" or "It's a hoax" objection.

Abstraction. Scientific thinking, Arendt observes, unfolds in a realm outside the political domain, independently of the political processes and values of the lifeworld. If one embraces those processes and values, one might be suspicious of something whose powerful force operates independently of these. This is the "I am not a scientist" or the "Scientists are their own special interest group" objection.

Open-ended. As Arendt says, quoting Schrödinger, nature acts so differently from what we observe in our normal environment that "*no* model shaped after our large-scale experiences can ever be 'true' " or complete,[50] in what I have called the "jury is still out" objection. A contemporary illustration is former EPA head Scott Pruitt's reason for not taking action about global warming: "So I think there's assumptions made that because the climate is warming, that that necessarily is a bad thing. Do we really know what the ideal surface temperature should be in the year 2100, in the year 2018? That's fairly arrogant for us to think that we know exactly what it should be in 2100."[51] This is Pope Urban's principle, updated.

Acting into Nature. Arendt's account emphasizes that the work of these institutions operating outside the processes and values of the lifeworld may introduce unforeseen dangers into the world. We can call this the "They don't know what they are doing" or the "Frankenstein" objection. Examples include efforts to block building heavy-ion accelerators for fear they would create black holes to destroy the universe.

Each of these vulnerabilities has been exploited by science deniers. Organizations have sprung up whose entire purpose is to promote distrust, not in scientific findings themselves, but in the institutions that create such findings. Such organizations can be called social Iagos, after the character from Shakespeare's *Othello* who seeks to advance his career by promoting suspicion.[52] This despicable practice seeks to undermine not just the authority of science, but of expertise itself. Those who seek to counter both social Iagos and others who attempt to either restrict the public space to the privileged or remove it entirely face ugly retaliation by watch lists and misinformation machines.[53]

Third, Arendt identified two key specific tactics that we see operating in science denial. One is to treat such facts as opinions, then oppose these opinions to others. This is possible because a fact is already a social thing; "it concerns events and circumstances in which many are involved; it is established by witnesses and depends upon testimony; it exists only to the extent that it is spoken about, even if it occurs in the domain of privacy."[54] This of course is especially true of scientific facts, for these are not just the result of testimony but products of institutions and bureaucracies with their own needs and interests. Their findings can thus be the outcome of compromises, and can conceal issue advocacy. In his book *The Honest Broker*, Roger Pielke Jr. uses the example of the famous "food guide pyramid" to illustrate this.[55] The very social character of facts, that is, can be used to discount them. Furthermore, the scientific infrastructure, and the production of scientific findings, tends to withdraw from public view, and therefore seem inconsequential, unfounded, or superfluous.

Finally, Arendt's writings provide some clues for what it would take to keep the public space open and nourish the compelling element that would have to underlie scientific authority. There is no quick fix. It would require developing a recognition of science as augmenting the elementary problem of humans living together, and of what is happening inside the workshops as responding to the world. It would require generating trust that the authority of science is augmenting the world, not exerting force on it. It would require a sense that loving the world means accepting that the numbers matter. It would require making humans realize that numbers can draw them closer to their earthly surroundings rather than only alienate them from it.

Arendt's way of addressing crises was generally to trace the story of how our tradition delivered us into them; such a story is necessary because it reads things from the inside rather than from the outside. Stories make things appear that you cannot see or think about otherwise. They also build their own canon for understanding crises, and thereby weave readers together into a community. Putting out our own story, making it an object for us all to contemplate, transforms it from a fleeting occurrence to the first step in what Arendt liked to call "reconciliation with reality." Telling such a story makes sure that, faced with an existential threat, humans have all the relevant information for their next step. It guarantees that members of *Homo sapiens* will decide in the action they take in response to that threat who and what kind of species—and how intelligent—that species really is. Telling such stories thereby reflects a commitment to what King calls Arendt's "ethic of pitiless responsibility for the world,"[56] and ensures that not only the teller but also the audience does not remain innocent.

CONCLUSION

But it's real for us!

It's real for us!

Doesn't matter what the muggles say,

It's real for us!

—LAUREN FAIRWEATHER

FOUR HUNDRED YEARS AGO, Francis Bacon came up with a terrific idea. Let's stop learning about nature accidentally. Each country needs to establish laboratories for trained people to investigate nature systematically. Some labs will study health, others weather, energy, communication, navigation, and the environment. The labs can collaborate on their findings, and the countries can use the findings to govern better. This will make human life flourish, as well as the fortunes of these countries.

Inspired by this vision, various nations began to train and support scientists, and the development of their procedures and instruments. Over the last century in particular, the United States and other nations have built up what is now in effect a global scientific workshop. This global workshop has made great contributions to human life, from providing an understanding of the origin and structure of the universe, to improving health standards and reducing air pollution. Great challenges still lie ahead, such as finding ways to keep epidemics, world hunger, clean water supplies, and climate change under control.

What went wrong? Despite all the progress, in the past few years in particular, more blatantly than ever, politicians and others are confident that they can ignore the findings of the scientific infrastructure.[1]

Lauren Fairweather's song "It's Real for Us" is about how a young person's love for the magical world of Harry Potter, and the belonging she feels in it alongside others like her, helps her cope with a world that she finds difficult and alien. Substituting magic for what's real sometimes functions, helping individuals pursue their desires and dreams. Turn this upside down, however, and you get the current science denial worldview of many politicians. Such people exhibit the same sentiment and tendency to magical thinking, though without Fairweather's self-conscious irony. Whether they actually believe in the magical world or are simply spinning things to get votes does not matter; what matters is that their substitution of science denial myths and cherry-picked or fake facts works for them and their voters.

Traditional approaches to countering denial do not work because they generally address specific acts of science denial rather than the dynamic that encourages such acts. That is playing whack-a-mole, because instances of science denial will only continue to crop up elsewhere. Science deniers are historical pawns rather than movers and shakers, pawns less even of special interests like the fossil fuel industry than of larger forces, brought to light by the writers in this book, that have to do with science itself. That is why denouncing, moralizing, conducting exposés, and doing epistemology have such little effect. *Denouncing* science deniers addresses only specific people, politicians, or claims, and leaves intact the social and political atmosphere in which they can get away with it. *Moralizing* only makes the moralizers feel superior. *Exposés* are easily ignored and can be accused of being tainted. Conducting *epistemology*—proclaiming something like "Science works!"—preaches to the converted and comes off as aloof and abstract.

Hoping for politicians with integrity is wistful thinking. One must start by understanding what makes the social and political atmosphere in which science denial takes place flourish, and what can be done about it.

The person in the above photograph appears comfortable. To confront science denial we have to understand what stories are unfolding in his head and where they came from. One serious problem is the huge disconnect between our perspective and his, which makes the whack-a-mole approach tempting. Traditional methods don't overcome the disconnect. They only bonk him on the head. They don't get inside it. If this person's stories don't change, his science denial will pop right up again.

The scholars and authors discussed in this book can help understand the dynamic of science denial, and what has to be done to counter it. Part of the dynamic is that the very structure of science creates vulnerabilities. As I noted in the Introduction, Facebook is a good analogy here; the very gears that make Facebook a terrific technology and socially wonderful—it allows us to connect and share—are the same ones that facilitate trolling, the flourishing of hate groups, the dissemination of fake news, dirty political tricks, and the subversion of privacy. Similarly, the same features that make science work also make it vulnerable to science denial. The authors just discussed allow us to identify several features that are strengths of science: it's a collective enterprise (Bacon, Weber, and Husserl), it's technical and abstract and requires special training (Galileo, Descartes, Husserl), and it's fallible (Galileo, Descartes). Furthermore, its power also comes from the fact that it can act into nature (Shelley, Arendt), can be passed on as a tool (Husserl), and has social and cultural consequences (Galileo, Vico, Comte, Weber, the Ottoman experience, Husserl).

But these six features can also turn into weaknesses that fuel science denial. That it's a collective means that it can potentially promote elite or disguised interests and amount to a "hoax." That it's technical and abstract can make legitimate people dismiss it, saying, "I am not a scientist." That it's fallible can appear to make it reasonable to say that "The jury is still out." The fact that it acts into nature can expose scientific projects to fears of producing Frankensteins. That its tools can be easily passed on means that their users can neglect or forget what is required to maintain them. That science has social and cultural consequences—including threatening deeply held beliefs—can make it seem to threaten genuine human values.

Someone is bound to object that it is hard to draw the line between the strengths and weaknesses of science. How can you tell, for instance, when a scientific collective is legitimate or pursuing an agenda, or when a model is solidly grounded enough to act on and when action is premature? I've often heard science deniers claim, for example, that Galileo defied the consensus of his time and was right. The answer lies in the motives of those who question the scientific consensus. Is their primary interest to understand nature using the best available scientific tools, improving them if there is doubt, or is it to find some wedge to justify a social or political agenda? Two key differences between Galileo and the renegade scientists championed by science deniers are that the consensus that Galileo defied was not a scientific but a political one, and that Galileo had evidence that he was willing to show and discuss with anyone interested. Modern-day science deniers are generally like Cremonini and refuse to look through the telescope.

Like reducing crime, improving the authority of science requires both short-term tactics and long-term strategies.

SHORT-TERM TACTICS

Short-term tactics add things to the existing public space in an attempt to deter and interfere with science denial. They attempt to fix the public space as it is. Here are five short-term tactics, each practiced by at least one of the figures in the preceding story.

1. DEMAND RESPONSIBILITY: PLEDGES

It is easy, in the abstract, to describe the difference between good and bad politicians. The former ground their decisions in facts, based on evidence and inquiry into how the world works, while the latter advance platforms that describe how they would like the world to behave. Good politicians understand the possible and seek it through fact-based means; bad politicians promote the pleasing and entice us through fantasies based on falsehoods and magical thinking. In practice, it can be difficult to tell the difference, but pledges such as the ones described in the chapter on the Ottoman experience can help, forcing politicians to shoulder responsibility for their positions.

Opponents of pledges say such statements are pointless and that "I pledge to uphold the Constitution" should suffice. They have a point. Pledges inhibit elected officials from compromising, negotiating, and changing their minds, which wise leaders sometimes do, and generally have more to do with asserting ideological purity than the realities of governing. But a properly crafted pledge can improve voters' ability to tell the difference between authentic and spurious leaders. Consider this: "I pledge to defend and maintain the scientific infrastructure of the country, and to let my decision-making be guided by facts rather than ideology or financial interest." That's reasonable and open-ended, because those who let gut instinct, ideology, class, or personal interest determine how the world works do not act in the public and national interest. How politicians view the scientific process—first diagnosing a problem, then investigating it and examining evidence, and only then moving to a solution—tends to reflect how they approach other issues. Pledges can also help voters determine whether a politician is really fully behind a statement or merely adopting it to get elected.

2. EXPOSE HYPOCRISY: GOING BACK AT THEM

Another tactic is to show how science deniers betray the very values they profess, and throw that fact back in their faces. It should not be necessary to make the incendiary comparison, made at the end of the chapter on the Ottoman experience, between ISIS militants and certain science-denying Washington politicians. But citing facts and honing arguments has not worked. Strong, documented editorials have little effect.[2] The reason is that Washington politicians believe themselves motivated by a higher authority. They see opposition to their actions not as being grounded in reality but as the confrontation of another (inferior) ideology with their own, far superior fundamentalist or libertarian one. Some wilier and more provocative way thus needs to be found—via startling metaphors, for instance—to expose the dishonesty of these politicians and how they are betraying the values of the country they profess to honor. Calling politicians the same as people they say they hate is not designed to win them over, though it may make them more careful about their claims. Still, a modern-day Galileo would be even more confrontational, and bluntly call out science deniers, "You are the anti-American ones!"

So back to the comparison between ISIS militants and science-denying US politicians: both believe that they are motivated by higher authority and that mainstream culture threatens their beliefs, and both want to damage the means by which that mainstream culture survives and flourishes. Is that such an over-the-top comparison? True, destroying objects is not in the same category as interfering with ideas. People whose other actions include hacking off the hands and heads of the innocent cannot be compared with elected officials tinkering with legislation. But how far over the top is it? When the North Carolina state legislators forbade incorporating scientific findings into state policies by state law, it damaged the ability of the state's officials to protect its coastline, its resources, and its citizens; it prevented other officials from fulfilling their duty to advise and protect innocent citizens against threats to life and property.

This and similar instances are not a matter of politicians disagreeing about whether and how to address threats like sea level rise and climate change. That's a political issue. Instead, it is a case of politicians deliberately destroying our ability to collect and use information about such threats. Such politicians are essentially saying: "Our beliefs are enough, our goals sufficient justification for our actions; we don't pay attention to the scientific findings and we'll make sure you can't, either." This is a moral issue; it is ideologically motivated cultural vandalism.

At debates and press conferences, such politicians should be asked: "Explain the moral difference between ISIS militants who attack cultural treasures and politicians who attack the scientific process." How they respond will reveal much about their values and integrity.

3. USE COMEDY AND RIDICULE: BE SCATHING, PROVOCATIVE, AND INFLAMMATORY

The magician James Randi once exposed a popular televangelist by playing recordings of secret transmissions between an audience plant and the televangelist. The televangelist declared bankruptcy the next year.

Doonesbury © *2011 G. B. Trudeau. Reprinted with permission of ANDREWS McMEEL SYNDICATION. All rights reserved.*

The incriminating evidence against science denial is rarely as direct and dramatic; science deniers muddy the waters with cherry-picked data, fake experts, and uncertainty. But comedy is often as effective in revealing the dynamics. Comedians have an ability to speak truths—truths that people are afraid to talk about, that people can't even see—in a way that breaks through resistance, cuts through codes, and speaks truth to power. They have a license to be inappropriate. A Doonesbury cartoon strip once featured an "honest" science denier interviewed on a radio talk show. "I don't oppose sound climate policy because it's flawed," he says. "I oppose it because I care *much* more about my short-term economic interests than the future of the damn planet. *Hello*?" A comedian's ability to be transparent and say unpleasant truths in a funny, satirical way invites trust: this is one reason why a Pew Research poll of public trust of news sources ranked the *Daily Show* higher than the *Economist*. Comedy can also expose opportunism masking as skeptical science.

Entire web pages are devoted to mocking climate deniers and their reasoning, or lack of it. A quick sample: How many climate deniers does it take to screw in a light bulb? None—it's more cost-effective to live in the dark!" Or: "It's too early to see if it needs changing!" This tactic exploits the fact that jokes can be disruptive by being both accessible and illuminating. Other examples of effective humor targeting science denial are the *Magic School Bus* spoof of anti-vaxxers, Jimmy Kimmel on the same topic, and numerous John Oliver and *Saturday Night Live* episodes.

These shows not only expose hypocrisy but provide viewers with a language for understanding what is happening, putting science denial in a wide enough context that the manipulation becomes visible. Humor contributes to what the American philosopher C. S. Peirce called "the social impulse" that disrupts "tenacity," or the urge to cling to select beliefs, by drawing listeners into a wider and wilder space in which the presence of more factors comes into play. In Weber's terms, what such humor does is to illustrate in concrete detail that science deniers are adopting the "ethics of conviction" as opposed to the "ethics of responsibility."

4. TELL PARABLES: METAPHORS AND FABLES

A fourth strategy is to tell parables involving science denial. I know "parable" has an old-fashioned flavor that seems to indicate that there's an embedded moral, but this is exactly what I mean. A parable, like an Aesop's fable, is a real or fictional story with a built-in, easily graspable lesson. It provides an effective teaching approach. After all, most people learn more easily through stories than data. *Jaws* and *Enemy of the People* are good examples. These powerful parables expose the all-too-rational calculus of science denial. We need twenty-first-century Aesops to tell more dramatic stories of what happens when we wish away sharks.

Did you hear the one about the person who was convinced, not altogether wrongly, that the medical establishment was corrupt, and decided that he was the only person who could fix it? "Make America Healthy Again!" was his slogan. His campaign to be the next person in charge succeeded. His solution was to get rid of medical tests and lab tests, destroy thermometers for taking temperature and stethoscopes for detecting heartbeat. The people ended up worse off but happier, convinced that they were in good hands.

Provocative metaphors include comparing science deniers to leeches or viruses, who live off the world without contributing to it; stickers might be made that say "DANGER: POLITICAL VIRUS," accompanied by a picture of a virus, that might be stuck on campaign posters of science-denying candidates. Another is ostriches: when Congress effectively commanded the Centers for Disease Control not to examine handgun violence, a major cause of deaths in the United States, it amounted to ostrich-like behavior.

Developing and publicizing such parables calls for "science critics" who constantly challenge inauthenticity, ridicule pretense, and expose those who speak in code. Science critics would help match the impedance between the abstract world of scientific findings and the lifeworld. I do not mean conventional science writing, but writing that shows that scientific findings do not occur in abstract spacetime, but are beacons on a turbulent sea. This kind of writing will help invest scientific work with a powerful claim to authority.

5. PROSECUTE

A final strategy is to prosecute science deniers. In 2015, US senator Sheldon Whitehouse of Rhode Island proposed that organizations bankrolling campaigns of climate science disinformation should be investigated for possible violation of Federal law. The law in question prohibits "racketeering"—a fraudulent business activity that includes conspiracy to deceive the public about such things as risk. Such laws have been successfully used to prosecute tobacco companies for misleading the public about the hazards of smoking.

That's a great idea. What's the difference between endangering the public by hiding evidence that smoking is hazardous, and endangering the public by concealing evidence of climate change? The crime is like shouting "Stay put! Everything's OK!" in a burning store so that people carry on shopping. Some might say that prosecuting science deniers is censorship and a denial of free speech, but if being misleading and deceptive about hazards isn't a crime, it should be.

We should legally target those who seek to block scientific information from being used to protect life and property. With the displacement of people due to global warming already starting, we need to prosecute people who disrupt our ability to use the knowledge we have to develop solutions. They should be forced to pay for the damages, personal and financial.

This tactic changes the context so that science denial becomes irrational, leading to lawsuits and heavy costs. It creates situations in which individuals and companies become legally responsible for actions that require the incorporation of science into decisions. An example is the New York state attorney general investigating Exxon for lying about climate change. It is true that science deniers are well organized and well funded, and are able to drag out court cases for years. They do not hesitate to use media connections to play the free-speech card, and they accuse adversaries of witch hunting. Still, it is right to have laws on the books against withholding information about the means to protect life and property. Avoiding prosecuting powerful people just because they are wealthy and well connected may be difficult, but it is not wrongful. Furthermore, a well-publicized court case, even a losing one, can exert positive cultural force; the Scopes trial is a good example.

THESE FIVE SHORT-TERM TACTICS cannot promise fundamental changes to the existing social world that we've inherited. However, they can discourage lazy and ideological thinking, curb the human appetite for fake assertions, and entice citizens to look past private interests and to regain an appreciation for the natural world. They seek to disrupt clinging to ideas past their sell-by date, not by imposing an ideology, but by increasing the damaging consequences for magical thinking in an environment that encourages it, and increasing as well the incentives for being truthful. These tactics involve taking aggressive steps. But explaining the importance of science in addressing crises over and over again has not been sufficient. Fighting science denial is not just for scientists and educators, but for lawyers, comedians, storytellers and other citizens. Science deniers need to be effectively called out for irresponsibility and for betraying values, and even for the illegality of their behavior. These five tactics will not eradicate science denial. But doing all of them all the time may help discourage politicians who practice it from getting elected.

LONG-TERM STRATEGIES

The authors discussed in this book also had their own era-appropriate suggestions for long-term strategies to transform the public space.

1. MAKE SADDLE BURRS

One way is for scientists and educators to continue to act in that public space, to practice science and articulate its relevance for the lifeworld—to keep being infuriatingly stubborn and creating work that acts like burrs under the saddles of science deniers, making their ride difficult. In apolitical contexts, occasionally simply repeating the science often enough works. One example is the way that publication and publicity surrounding the scientific analysis of the "sliding stones" of Death Valley—allegedly moved by aliens but actually by natural causes—was able to curb pseudoscientific accounts. Another is the long-term impact of anti-pseudoscience crusaders such as Carl Sagan and Martin Gardner. Evidence of the impact of science, and of neglecting it, should also be made visible in the lifeworld. Once every five years, legislators of every country above a certain grade should be forced to have a retreat at the Mer de Glace, where the fact and danger of global warming is a visceral experience.

2. RECALL AND RETELL

Recall and retell stories about past action in that space; the triumphs and the failures of science in public, and tales of bygone "infuriating stubbornness." Of the Flint, Michigan water crisis that began in 2014, for instance, or the Elk River chemical spill in West Virginia that same year. The motive is not nostalgia, academic scholarship, or glorification of the scientists but a means to renew scientific authority without embracing it as a "cause."

3. COOPERATIVE NETWORKS

Where possible, encourage local yet interconnected cooperative networks such as the civic webs envisioned by Jefferson in his ward system, or the political associations that de Tocqueville noted served to draw American citizens' attention past their immediate interests to larger issues. This exposes citizens to things they do not want or expect, and engages them in collaborative problem solving.

4. ADDRESS CURRENT EVENTS

Seize any opportunity to address current events with a scientific-technological dimension—the way Arendt did Sputnik, the Holocaust, and other controversies—in which the issues are fresh in people's minds and traditional assumptions are up in the air. Senator Inhofe's snowball can be used to illustrate the difference between climate and weather, Bermdoggle the science of beaches, and the water catastrophes in Flint,

Michigan, and the Elk River in West Virginia the need for laboratory analyses of polluted water.

5. TELL THE STORY

Tell the story of how we got into this situation. Science denial is like what Arendt said about totalitarianism; while historically unprecedented, it arose because of the way our traditions developed. The longest-term strategy is to keep doing for science denial what she did for totalitarianism—keep telling the story of what led to it. This story would include how people promoted the idea of the workshop, defended its authority, and defended as well the special training required of those who work in it. The story would include how other people came to point out the dangers and vulnerabilities of the workshop, and to suggest some ways to counteract these. The story, in short, would have to be a mirror in which each actor—workshop participant, nonworkshop participant, science denier—could recognize themselves and other participants. It would have to highlight the difference between individual acts of science denial and the atmosphere that makes them possible; the difference between moles and the machine. It would have to exhibit, not hide, the vulnerabilities of science, or what drives the moles. The details make the story fun and compelling; the seriousness makes following it worthwhile.

Finally, it would have to be a motivating story, one that does not let readers off the hook regarding what comes next. The story needs the drama of telling people on a boat that there is a storm, the boat is heading toward a dangerous reef, there are tools around to get back on course, but those in charge are ignoring the advice of the navigators, discarding the navigational aids, and thus endangering everyone aboard. The story would awaken the existential sense that humans decide by their actions what kind of creatures they are—whether they can face a crisis that they themselves have caused, or will let themselves slide, as Vico depicted, into collapse. The story has to make us realize that our decisions about science depend on who we are and who we are to become. We certainly have enough material to write this story. For science denial affects public health, the welfare of future generations, and the fate of the planet. We have the "motive and cue" to respond to the story, in Hamlet's words, as well as the "reason and capability." How we react reveals how we are living out our inheritance—the workshop we have built—and therefore who we human beings are.

LET ME END with a little parable. Suppose I were responsible for taking you and your family on a cross-country trip in a bus, and suppose you learned that the consensus of mechanics was that my bus was unsafe. I dismissed what those mechanics said, telling you that they were all wrong and engaged in a conspiracy for their own benefit—after all, they are getting paid to be mechanics! Suppose I then threw out the bus's spare tires and jacks, and switched off my ability to contact roadside assistance. Suppose I then opened the bus door and said, "Get in!"

Would you be angry? What would you do?

Earthrise.

ACKNOWLEDGMENTS

The idea for this book occurred to me in the fall of 2014, after I was asked to give a TEDx talk at CERN, the international laboratory in Geneva, Switzerland. The theme of this wonderful event, associated with the sixtieth anniversary of the lab, was "idea boats," symbolized by little origami folded paper boats that were plentifully distributed throughout the auditorium. I was selected to be the first speaker, and the choice of topic was up to me. CERN is one of the crown jewels of the global scientific workshop, a triumph of humanity's attempt to understand nature. I thought I might use my fourteen allocated minutes to try to explain why some politicians, and the people who vote for them, could possibly reject the findings of this workshop, and what could be done about it. But I was unable to condense my reasoning into fourteen minutes. Among other mistakes, I forgot to show the slide I had of a menacing shark of the sort seen in the movie *Jaws*, symbolizing dangers that you cannot avoid without recourse to science. A few minutes into the talk I finally remembered to show that slide, and clicked the remote so that the image of a terrifying shark suddenly appeared on the screen behind me. A reporter covering the event, whose name I cannot remember, wittily tweeted, "We're going to need a bigger idea boat!" This book is my full-scale version of that bigger idea boat. My first debt is therefore to James Gilles and Claudia Marcelloni of CERN, who extended the invitation to me.

Like any boat, this one was partly created from already existing pieces. Many passages originated in my column in *Physics World*, called "Critical Point," which I've written for nineteen years. Published by the Institute of Physics, *Physics World* is a superb, entertaining, and worldly magazine, edited by Matin Durrani. I owe an enormous debt to Matin for having invited me to write the column in the first place, as well as for his outstanding editing and for allowing me to explore a wide range of topics in my columns. Several of these columns are incorporated here. Chapters that contain some material that first appeared in these columns include Chapters 1 ("Cooking Bacon," November 2015, and "Diversifying Utopia," March 2016), 2 ("Entry Denied," May 2017), 3 ("In Praise of Descartes," August 2016), 5 ("Franken-Physics," December 2016), 6 (Storytelling Matters," April 2016), 7 ("Why Don't They Listen?" May 2014), and the Conclusion ("Fighting Science Denial," September 2016, and "This Time It's Different" January 2017). I am grateful for permission to revise and adapt this material.

This book, I think, is an example of the role that the humanities can and ought to play in preserving cultural health in a scientifically and technologically permeated world. I could not have written this without considerable help from others in the humanities who read and commented on parts. David Dilworth commented on the Introduction, while Joseph D. Martin, Peter Pesic, and Robert C. Scharff read the material on Bacon. Peter Carravetta, Arianne Nicole Margolin, and Mark Peterson helped me with the chapter on Galileo, Margolin being the one who drew my attention to the burning geometry books in the Le Sueur painting. Scharff and Andrew Platt read the Descartes chapter, Carravetta the Vico chapter, and Peter Manning and Elyse Graham the Shelley chapter. Scharff, the most insightful philosophical commentator on Auguste Comte, first drew my attention to the importance of that thinker; Mary Pickering,

Comte's biographer, discussed his life with me; and Daniel Labreure showed me around the Comte museum. Linda Catalano, an exceptionally thoughtful reader, insightful critic, and passionate teacher, provided suggestions on the Weber chapter, while Scharff and Anthony Steinbock read the Husserl chapter, and Anne O'Byrne and Phillip Nelson read the Arendt chapter. I am grateful to Lauren Fairweather for allowing me to cite her lyrics to her song "It's Real For Us." Delicia Kamins helped take the picture of Dashiell and me on the train tracks. I am, of course, responsible for any mistakes that slipped past these diligent scholars. I would like to thank Robert C. Scharff in particular, whose penchant for returning material sent to him for commentary with an extensive use of Track Changes to provide markups and elaborate commentary has given rise to a new adjective among his interlocutors: "scharffing."

Jennifer Gaffney's invitation to speak at the Arendt Circle gave me an opportunity to run the Arendt material past scholars there and I am grateful for permission to reprint passages from that talk, published in the first issue of *Arendt Studies* (vol. 1, issue 1, 2017, 43–60). Eoin Gill and Sheila Donegan's invitation to speak at the Robert Boyle Summer School allowed me to present the freshly written introduction. Alissa Betz, the Philosophy Department's unbelievably skilled and steady ATC (Assistant to the Chair), made it possible to write this book while also serving as an administrator. Charles C. Mann, a longtime friend and outstanding writer, was always available for help and inspiration when I got tired, frustrated, or stuck. Edward S. Casey was also willing to listen and comment on my ideas. I also thank my students in several sections of PHI 112, 268, and 618 at Stony Brook University, who tolerated my obsession with reading and discussing many of these underappreciated writers. My wife Stephanie Crease read virtually the entire book several times, making sure that it was clear and accessible. She was also motivating in her own way; with the sun setting and rain threatening, she gave me a fifteen-minute ultimatum for finding Comte's gravestone in the Père Lachaise cemetery in Paris, which sent me down to the wire. I'd also like to thank my son Alexander, who came up with the idea of counting the stairs down to the Mer de Glace, and my daughter India, who came up with the idea of sailing over it in a paraglider. My dog Dashiell was always ready to take me for a walk after I exhausted the patience of everyone else.

I am indebted to Maria Guarnaschelli, who commissioned this book trusting that I would make something more of it than was evident in the proposal, and to Quynh Do, who took over editing the book from Maria. Quynh read the manuscript more carefully than any book editor I ever had, and I learned much from her comments. I am also indebted to Norton's project editors Amy Medeiros and Dassi Zeidel, to production manager Lauren Abbate, and to copyeditor Gary Von Euer.

I dimly recall once hearing William H. Gass say something to the effect that anger is an important and maybe even necessary motive for finishing a book. I think I know what he means. Anger of two sorts possessed me throughout the writing process here. One was anger at politicians who pontificate about American values—and even Christian values—and then act in self-interested ways that benefit the top 1 percent and destroy the country and the planet for the rest of us. The other source of anger was at certain colleagues in the humanities who regard interacting with the sciences as something that is "not done" by real humanities scholars—as selling out and "service" at best. But in a world permeated by science and technology, humanities scholars have to engage the sciences as a part of the fundamental duty of the humanities to connect. I owe a final and special sort of debt to these two groups of people, without whom this book might have been conceived but would have taken much longer to finish.

NOTES

INTRODUCTION

1. EDYTEM is an acronym for "Environments, Dynamics and Territories of the Mountain," an interdisciplinary laboratory organized to study environmental and social issues of mountainous regions that is a joint research project of the Savoie Mont Blanc University and the National Center for Scientific Research of France.
2. A quick review of the workshop findings: The atmosphere is warming; here, for instance, is a website maintained by the National Aeronautics and Space Administration's Goddard Institute for Space Studies, regarding the Surface Temperature Analysis (GISTEMP), an estimate of global surface temperature change: https://data.giss.nasa.gov/gistemp/ . The reasons why the workshop experts think that this warming is human-caused are clearly outlined by Columbia University astronomer David J. Helfand at the following website: http://www.indiana.edu/~ensiweb/Main%20Source%20of%20CO2%20Narrative.pdf. A quantification of the consensus among the workshop experts can be found in the following scientific paper: http://iopscience.iop.org/article/10.1088/1748–9326/8/2/024024/meta.
3. Inhofe, *The Greatest Hoax: How the Global Warming Conspiracy Threatens Your Future* (Washington, DC: WND Books, 2012). Inhofe, a US senator (R-OK), has had an enormous impact on US climate policy as chairman of the Environment and Public Works Committee. See also Donald J. Trump's numerous tweets on the subject of global warming, including, "Ice storm rolls from Texas to Tennessee—I'm in Los Angeles and it's freezing. Global warming is a total, and very expensive, hoax!" Donald J. Trump, 12/6/2013, Twitter. "The concept of global warming was created by and for the Chinese in order to make US manufacturing non-competitive." Donald J. Trump, 11/6/2012, Twitter.
4. Senate Minority Leader Mitch McConnell (R-KY), asked if greenhouse gas emissions cause global warming (10/2/2014), said, "I am not a scientist. I'm interested in protecting Kentucky's economy." https://www.courier-journal.com/story/news/politics/elections/kentucky/2014/10/02/mcconnell-climate-change-scientist/16600873/. Florida governor Rick Scott, when asked about global warming, said, "I am not a scientist," http://miamiherald.typepad.com/nakedpolitics/2014/05/rick-scott-wont-say-if-he-thinks-man-made-climate-change-is-real-significant.html.

5. Former EPA head Scott Pruitt (2/7/2018): "So I think there's assumptions made that because the climate is warming, that that necessarily is a bad thing. Do we really know what the ideal surface temperature should be in the year 2100, in the year 2018? That's fairly arrogant for us to think that we know exactly what it should be in 2100," washingtonpost.com/news/energy-environment/wp/2018/02/07/scott-pruitt-asks-if-global-warming-necessarily-is-a-bad-thing/?utm_term=.cf948b6201f2. Scott Pruitt (1/18/2017): "Science tells us that the climate is changing, and human activity in some manner impacts that change. The human ability to measure with precision the extent of that impact is subject to continuing debate and dialogue, as well they should be," https://www.c-span.org/video/?421719–1/epa-nominee-scott-pruitt-testifies-confirmation-hearing&start=2040. Former Texas governor Rick Perry: "Well, I do agree that there is—the science is not settled on this. The idea that we would put Americans' economy at jeopardy based on scientific theory that's not settled yet, to me, is just nonsense," https://www.nytimes.com/2011/09/08/us/politics/08republican-debate-text.html. Perry had an enormous impact on US energy policy after becoming US Secretary of Energy.
6. A classic example is the "Food Pyramid" described by the US Department of Agriculture in 1992 as a way to identify healthy diets, which was the product of much internal political negotiation. A brief description of the process can be found in Roger A. Pielke Jr., *The Honest Broker: Making Sense of Science in Policy and Politics* (Cambridge: Cambridge University Press, 2007).
7. For the case of global warming, see Spencer Weart, *The Discovery of Global Warming* (Cambridge, MA: Harvard University Press, 2008), or online at https://history.aip.org/climate/index.htm.
8. Weart's book cited in the previous footnote illustrates this as well.

CHAPTER ONE: FRANCIS BACON'S NEW ATLANTIS

1. York House itself was demolished a few decades after Bacon's birth, but the Watergate was left intact as a public monument. In the nineteenth century an embankment built in the river left it landlocked.
2. Its ruins are still visible, accessible by a lovely two-hour walk outside the city of St. Albans through quiet fields and pastures. The ruins consist mainly of brick and stone walls that once surrounded the chapel, great hall, and dining room. Its most dramatic remaining part is the porch into the Great Hall, with features such as Roman figurines, a Latin inscription, and the Royal Arms of Queen Elizabeth I, who had appointed Nicholas.
3. Today, much of Gray's Inn looks the way it did in Bacon's time. A visitor walking through Gray's Inn Square can find the four-story buildings on the west side, now offices, which were the lodgings first of Nicholas and then of Francis. On the south side of this square is the Great Hall, where Francis's plays were performed. See Robert P. Crease, "The Physical Tourist: Francis Bacon's London," in *Physics in Perspective* 19 (2017), 291–306, DOI 10.1007/s00016–017–0207–6, http://rdcu.be/uVZz.
4. Francis Bacon, "A Device for the Gray's Inn Revels," in B. Vickers, ed., *Francis Bacon: A Critical Edition of the Major Works* (New York: Oxford University Press, 1996), 54–55.
5. Vannevar Bush, Science: The Endless Frontier (A Report to the President by Vannevar Bush, Director of the Office of Scientific Research and Development, United States Government Printing Office, Washington, 1945), https://www.nsf.gov/od/lpa/nsf50/vbush1945.htm.

6. Bacon uses this image in his essay "The Interpretation of Nature," in *The Advancement of Learning*, and elsewhere. Likening nature to a book was a brilliant rhetorical stroke, because it goes far back in Christian tradition. See for instance Peter Harrison, *The Bible, Protestantism, and the Rise of Modern Science* (Cambridge: Cambridge University Press, 2001).
7. Nieves Mathews, *Francis Bacon: The History of a Character Assassination* (New Haven, CT: Yale University Press, 1996), 233.
8. Bacon, *Advancement of Learning* I, IV, 2.
9. Francis Bacon, *The Philosophical Works of Francis Bacon*, ed. J. Spedding (London: Longman, 1861), 4:32 (in "The Plan of the Work"). Part 1 was a revised version of the *Advancement of Learning*. Part II was new: the *Novum Organum*, a logic or set of instructions for investigating nature. The title alluded to Aristotle, whose *Organon* contained a logic for reasoning but which Bacon considered too concerned with proofs and arguments. Bacon's radical revision was a logic of discovery, presented in two books of aphorisms: short, comprehensible condensations of his ideas.
10. Each of these is mentioned in Mathews, *Francis Bacon*. How many other public figures can say that someone has posted their worst jokes online? http://www.telegraph.co.uk/culture/culturenews/7719397/Sir-Francis-Bacons-bad-jokes-go-online.html.
11. Mathews, *Francis Bacon*.
12. The birth of the Anthropocene, the proposed era when human beings left their mark on planetary processes, is sometimes pinned to the date when the level of CO_2 in the atmosphere was at its lowest point. That was 1610, in Bacon's lifetime. *Nature* 519, 171–80 (March 12, 2015), doi:10.1038/nature14258.
13. Sandra Harding, *Whose Science? Whose Knowledge?* (Ithaca, NY: Cornell University Press, 1991), 43.
14. Carolyn Merchant, *Death of Nature: Women, Ecology, and the Scientific Revolution* (New York: Harper & Row, 1980), 168.
15. Peter Pesic, "Francis Bacon, Violence, and the Motion of Liberty: The Aristotelian Background," *Journal for the History of Ideas* 75 (2014): 69–90; "Proteus Rebound: Reconsidering the 'Torture of Nature,' " *Isis* 99, 304–17 (2008); "Wrestling with Proteus: Bacon and the ' "Torture' of Nature," *Isis* 90, 81–94 (1999).
16. Aubrey, *Brief Lives* (New York: Penguin Classics, 2000).
17. S. Weinberg, *To Explain the World* (New York: Harper, 2015), 202. Some others criticize Bacon for having no appreciation for the role of mathematics in science, and for suggesting that making discoveries is simply a matter of setting up the right conditions for observing nature.

CHAPTER TWO: GALILEO GALILEI AND THE AUTHORITY OF SCIENCE

1. "The Preaching of St. Paul," painted in 1649 or six years after Galileo's death, was clearly part of a Counter-Reformation effort to combat Galileo's influence.
2. Mark Twain, *Innocents Abroad* (New York: Library of America, 1985). The Duomo's pendulum was replaced in 1587, so Twain saw the offspring of the Abraham pendulum.
3. This episode is recounted in Mark Peterson, *Galileo's Muse* (Cambridge, MA: Harvard University Press, 2011), ch. 10.
4. Galileo titled them "Two Lectures to the Florentine Academy on the Shape, Location and Size of Dante's *Inferno*."

5. Peterson, *Galileo's Muse*, 229.
6. Alfred Whitehead, *Science and the Modern World* (New York: Macmillan, 1925), 2.
7. Galileo Galilei, *Sidereus Nuncius*, or *The Sidereal Messenger*, tr. Albert Van Helden (Chicago: University of Chicago Press, 1989).
8. Galileo, Letter to Kepler, in Stillman Drake, *Galileo At Work* (Chicago: University of Chicago Press, 1978), 162. Today we would not mention asps blocking their ears to keep from hearing, but ostriches burying their heads in the sand. Both legends are untrue; ostriches do use their heads to dig holes in the sand, but to make nests for their eggs rather than to blind themselves.
9. David Wootton, *The Invention of Science: A New History of the Scientific Revolution* (New York: Harper, 2016), 185–86.
10. In that respect its intended effect resembles that of the play-within-the-play of *Hamlet*, which the prince puts on "to catch the conscience of the King." This play sought to catch the intelligence of any sensible reader of Italian.
11. Galileo Galilei, *Galileo on the World Systems: A New Abridged Translation and Guide*, tr. M. A. Finocchiaro (Oakland: University of California Press, 1997), 199.
12. Galileo, *World Systems*, 307.
13. At one point in the book, for instance, Simplicio fears that without the guidance of Aristotle, or some other venerable authority, he will be lost—and Salviati replies that he will find guidance with "reasons and demonstrations (yours or Aristotle's) and not with textual passages or mere authorities because our discussions are about the sensible world and not about a world on paper." Galileo, *World Systems*, 127–28.
14. Depending on which calendar system you use, for Europe changed its calendar system from Julian to Gregorian about this time.

CHAPTER THREE: RENÉ DESCARTES

1. This and other biographical information is from Geneviève Rodis-Lewis, *Descartes: His Life and Thought*, tr. J. M. Todd (Ithaca, NY: Cornell University Press, 1999).
2. Letter to Mersenne, 11/13/1629, René Descartes, in *Oeuvres de Descartes*, ed. Ch. Adam & Paul Tannery (Paris: J. Vrin, 1897), 70.
3. *Dictionary of Scientific Biography*, ed. Charles Gillispie (New York: Scribner's, 1970), 4:52.
4. René Descartes to Mersenne, November 1633, in *Oeuvres et Lettres* (Paris: Gallimard, 1953), 947.
5. From Ovid, *Tristia*, "To live well you must live unseen." For Descartes, living well came first. Descartes to Mersenne, April 1634, in *Oeuvres et Lettres*, 951.
6. Rodis-Lewis, *Descartes*, 138.
7. From our enlightened position four hundred years later, it is easy to trivialize his struggle and say that he was "really" a Galilean and a closet atheist, and that his Catholicism was a protective pose. But this is fake scholarship and cheapens Descartes's struggle and those of many other European scholars in the same position.
8. The adjective comes from his Latin name Renatus Cartesius.
9. The law of refraction was certainly known to others before Descartes; it had been described earlier by Willebrord Snellius (1580–1626) and much earlier by the Muslim mathematician and physicist Ibn Sahl (c. 940–1000).
10. Rodis-Lewis, *Descartes*, 123.
11. Rodis-Lewis, *Descartes*, 118.
12. Rodis-Lewis, *Descartes*, 127.
13. If he seems to shift his views about things like doubt, error, and truth along the way, it's because he is describing what he learned about them in the process; it's similar to the

way pioneers often find themselves changing their minds about paths they have taken. Descartes is showing the reader how he found a path. He is map-making, not following a Google Map that has laid out the route ahead of time.

14. René Descartes, *Descartes' Conversation with Burman*, tr. and ed. J. Cottingham (Oxford: Clarendon Press, 1976), section 44, http://www.earlymoderntexts.com/assets/pdfs/descartes1648.pdf), 14. Isn't Descartes right? Doesn't our daily experience testify to the intermingling of mind and body? But it is difficult to explain with Descartes's own concepts. A careful scientist, he states something he knows to be true while refusing to conjecture how. It would be another three centuries before phenomenologists such as Edmund Husserl and Maurice Merleau-Ponty would give a better philosophical-scientific description of the mind-body relation. Besides, describing that relation is not what Descartes was really after, which was to show how the sequester was possible in preparation for scientific thinking. A thorough and solid examination of Descartes's remarks about what he called the "substantial union" of the body and the mind, and that it is an "immediate" and even "primary" datum of consciousness, is found in chapter 12 of L. J. Beck's *The Metaphysics of Descartes* (Oxford: Oxford University Press, 1965). Beck sums up Descartes's position in the following way: "The fact of [mind-body] interaction in itself is not a scientific fact, if science here means the physical sciences. On the side of the mind it is possible to give an analysis of the passions which can be well described as a phenomenology. On the physiological side, it is possible to work out an account of the bodily phenomena which occur 'à l'occasion' of the mental acts, or vice versa, to state even the corporeal conditions of consciousness. The fact of sensation is irreducible in itself. This does not create a problem of 'interaction'; it is a natural consequence of the fact of interaction" (275).
15. The book *Descartes' Bones* tells the interesting story in which Descartes's bones came to be treated as akin to relics, with his skull and one finger separated from the rest of the skeleton.
16. Russell Shorto, *Descartes' Bones: A Skeletal History of the Conflict Between Faith and Reason* (New York: Vintage, 2008).
17. Steven Weinberg, *To Explain the World: The Discovery of Modern Science* (New York: HarperCollins, 2015), 204–5.
18. Robert C. Scharff, *How History Matters to Philosophy* (New York: Routledge, 2014), 79.
19. His remarks about the force of everyday habit, the need to practice rigorous thinking, the way one can be dazzled by mathematical-like proofs, and his constant injunctions to be careful, pay attention, go step by step, and recheck frequently reveal a man painfully aware of the vulnerability to regression of workshop thinking. So does his statement, at the end of the *Discourse*, that if he hadn't had to fight to discover difficult truths—if he had been taught much of what he learned instead of learning it himself—he wouldn't have developed the habit of mind, that unbelievably fierce and independent character, needed to go beyond them.

PART II

1. The thought about how radical a change it was comes from Wootton, *Invention of Science*. Bacon, Galileo, and Descartes were not inventing it all themselves, of course. They were responding to powerful scientific and technological developments that had been under way since the late middle ages. Leonardo Da Vinci (1452–1519) had used scientific inquiry to help him paint nature and devise inventions. Andreas Vesalius (1514–1564) had dissected corpses prior to Descartes. Studies of blood flow by William Harvey (1578–1657) influenced Descartes. The vision of the workshop did not come out of the blue but had set

itself in motion, and had forced itself on their attention. When the philosopher Immanuel Kant (1724–1804) said that "Enlightenment is man's emergence from his self-imposed mental immaturity," he had this period in mind as humanity's adolescence.

CHAPTER FOUR: GIAMBATTISTA VICO

1. Hermann Kahn—an inspiration for the character Dr. Strangelove—was one of many defense analysts who regarded computers as the most effective tool for developing military strategy. Another was Kahn's friend Bernard Brodie, who wrote *A Guide to Naval Strategy* without having been on a ship or seen an ocean.
2. Isaiah Berlin, *Vico and Herder: Two Studies in the History of Ideas* (New York: Vintage, 1976), 3.
3. Giambattista Vico, *The Autobiography of Giambattista Vico*, tr. M. H. Fisch and T. G. Bergin (Ithaca, NY: Cornell University Press, 1944), 111.
4. H. P. Adams, *The Life and Writings of Giambattista Vico* (London: Allen & Unwin, 1935), 24.
5. Vico, *Autobiography*, 134.
6. Giambattista Vico, *On Humanistic Education (Six Inaugural Orations, 1699–1707)*, tr. G. A. Pinton and A. W. Shippee, intr. Donald Philipp Verene (Ithaca, NY: Cornell University Press, 1993).
7. Elio Gianturco, "Translator's Introduction" to Giambattista Vico, *On the Study Methods of Our Time*, trans. intr. Elio Gianturco (Ithaca, NY: Cornell University Press, 1990), xxii.
8. Vico, *On the Study Methods of Our Time*, 3–4.
9. Giambattista Vico, *On the Most Ancient Wisdom of the Italians*, trans. L. M. Palmer (Ithaca, NY: Cornell University Press, 1988), 45.
10. *Universal Right*, trans. and ed. Giorgio Pinton and Margaret Diehl (Amsterdam and Atlanta: Rodopi, 2000).
11. Later, Gentile's single attempt at publication was shown to be plagiarized and withdrawn, and he eventually committed suicide after being caught sexually abusing a servant.
12. Barbara Ann Naddeo, *Vico and Naples: The Urban Origins of Modern Social Theory* (Ithaca, NY: Cornell University Press, 2011).
13. Vico, *Autobiography*, 15.
14. While writing the *New Science*, Vico received a request to compose an autobiography, part of a then-novel project by a group of scholars partly inspired by Descartes's *Discourse*. But Vico realized that his own conclusions forbade him from writing as if he were a self-assured "I" who recounts what he did and suffered. "We shall not here feign what René Descartes craftily feigned as to the method of his studies simply in order to exalt his own philosophy and mathematics and degrade all the other studies included in divine and human erudition," Vico writes in his contribution. He continues, referring to himself in the third person: "Rather, with the candor proper to a historian, we shall narrate plainly and step by step the entire series of Vico's studies, in order that the proper and natural causes of his particular development as a man of letters may be known."
15. H. P. Adams, *Vico* (London: Allen & Unwin, 1935), 148.
16. Vico, *Autobiography*, 15.
17. Giambattista Vico, *The New Science*, tr. T. G. Bergin and M. H. Fisch (Ithaca, NY: Cornell University Press, 1968). I'll refer to the quotes by the paragraph number; this one is #331: "But in the night of thick darkness enveloping the earliest antiquity, so remote from ourselves, there shines the eternal and never failing light of a truth beyond all

question: that the world of civil society has certainly been made by men, and that its principles are therefore to be found within the modifications of our own human mind."

18. Vico, *New Science*, #142: "Common sense is judgment without reflection, shared by an entire class, and entire people, an entire nation, or the entire human race."

19. Vico, *New Science*, #414: "Human choice, by its nature most uncertain, is made certain and determined by the common sense of men with respect to human needs or utilities."

20. Vico, *New Science*, #341: "Man in the bestial state desires only his own welfare; having taken wife and begotten children, he desires his own welfare along with that of his family; having entered upon civil life, he desires his own welfare along with that of his city; when its rule is extended over several peoples, he desires his own welfare along with that of the nation; when the nations are united by wars, treaties of peace, alliances, and commerce, he desires his own welfare along with that of the entire human race. In all these circumstances man desires principally his own utility."

21. Vico, *New Science*, #218: "Men at first feel without perceiving, then they perceive with a troubled and agitated spirit, finally they reflect with a clear mind." #374: "Hence poetic wisdom, the first wisdom of the gentile world, must have begun with a metaphysics not rational and abstract like that of learned men now, but felt and imagined as that of these first men must have been, who, without power of ratiocination, were all robust sense and vigorous imagination."

22. Vico, *New Science*, #1108: "Men mean to gratify their bestial lust and abandon their offspring, and they inaugurate the chastity of marriage from which the families arise. The fathers mean to exercise without restraint their paternal power over their clients, and they subject them to the civil powers from which the cities arise. The reigning orders of nobles mean to abuse their lordly freedom over the plebeians, and they are obliged to submit to the laws which establish popular liberty. The free peoples mean to shake off the yoke of their laws, and they become subject to monarchs. The monarchs mean to strengthen their own positions by debasing their subjects with all the vices of dissoluteness, and they dispose them to endure slavery at the hands of stronger nations."

23. One might say the process is guided by Providence, but that is only a metaphor for the fact that humanity is constantly reinventing itself without planning ahead. Humans are locked up in history, unable to step outside it. Vico, *New Science*, #1108: "This world without doubt has issued from a mind often diverse, at times quite contrary, and always superior to the particular ends that men had proposed to themselves; which narrow ends, made means to serve wider ends, it has always employed to preserve the human race upon this earth."

24. Vico, *New Science*, #349: "Our Science therefore comes to describe at the same time an ideal eternal history traversed in time by the history of every nation in its rise, development, maturity, decline, and fall."

25. Whether Vico saw the *New Science* as possibly having an impact on this process is controversial in Vico scholarship; see Robert P. Crease, "Vico's 'Mirror Stage': Narrative, the *Scienza Nuova*, and the Barbarism of Reflection," in *Studies in 18th Century Culture* 24 (Baltimore: Johns Hopkins University Press, 1995), 107–19. It is true that Vico's position is not entirely consistent. He described humanity as developing through a causal cycle that takes a destructive turn. If humans saw this cycle, they would be horrified and take the required action to avoid its happening. But that action would have to come from some other resource, some other motivation, in human

nature that was not a part of the cycle whose description Vico left blank. I'll call this Vico's void. Later philosophers would fill it in.

26. Adams, *Vico*, 172.
27. Roberts's remark can be found at the Oyez website, a judicial archive of the US Supreme Court, under Gill v. Whitford: www.oyez.org/cases/2017/16–1161.

CHAPTER FIVE: MARY SHELLEY'S HIDEOUS IDEA

1. This is from Mary Shelley's introduction to the 1831 edition, in which she tells her version of the genesis of *Frankenstein*. In her biography *Mary Shelley*, Miranda Seymour points to the spins and embellishments that Shelley surely gave to her own creation story, such as the depiction of it occurring to her as a lightning bolt.
2. Thomas Moore, ed., *Letters and Journals of Lord Byron* (London: John Murray, 1830), chapter 14.
3. Hogg Memoir, *The Life of Percy Bysshe Shelley* (London: Moxon, 1858), 1:62, 71.
4. Anne K. Mellor, *Mary Shelley: Her Life, Her Fiction, Her Monsters* (New York: Routledge, 1988), 107.
5. Gary Harrison and William L. Gannon, "Victor Frankenstein's Institutional Review Board Proposal, 1790," *Science and Engineering Ethics* 21:1139, DOI 10.1007/s11948–014–9588-y.
6. Bruno Latour, *The Breakthrough Journal* (Winter 2012), https://thebreakthrough.org/index.php/journal/past-issues/issue-2/love-your-monsters.

CHAPTER SIX: AUGUSTE COMTE'S RELIGION OF HUMANITY

1. *Human Events*, May 31, 2005.
2. Andrew Wernick, *Auguste Comte and the Religion of Humanity* (Cambridge: Cambridge University Press, 2001), 5. Wernick provides the most thorough analysis of Comte's ideas about the religion of humanity. The best source for Comte's ideas about science is Robert Scharff, *Comte after Positivism* (Cambridge: Cambridge University Press, 1995). For Comte's life, see the thorough three-volume biography by Mary Pickering (Cambridge: Cambridge University Press, vols. 1, 2 (1993), vol. 3 (2009).
3. Pickering, *Comte*, 2:8.
4. Pickering, *Comte*, 3:580.
5. One of the first promoters of the word "positive" to mean accurate and precise as opposed to abstract and metaphysical knowledge was Madame Germaine de Staël. See Pickering, *Comte*, 1:65n16.
6. Pickering, *Comte*, 1:318.
7. Pickering, *Comte*, 1:371.
8. John Morley, "Comte," *Encyclopædia Britannica*, 13th ed. (New York: Encyclopædia Britannica, Inc. 1926), 816.
9. While the complete *Course* is available in French online, it is only available in a three-volume "freely translated and condensed" English version, *The Positive Philosophy of Auguste Comte*, trans. Harriet Martineau (London: Trübner, 1875), also online.
10. Scharff, *How History Matters*, 115.
11. "The supernatural agents are replaced by abstract forces, real entities or personified abstractions, inherent in the different beings of the world," Comte wrote. Auguste Comte, "The Nature and Importance of the Positive Philosophy," in R. Scharff and V. Dusek, eds., *Philosophy of Technology*, 2nd ed. (West Sussex: Wiley, 2014), 54.
12. "Nature and Importance," 55.

13. "Without the attractive chimeras of astrology," Comte writes, "or the powerful deceptions of alchemy, for example, where should we have found the perseverance and ardor necessary for collecting the long series of observations and experiments which later on served as a basis for the first positive theories of these two classes of phenomena?" Comte, "Nature and Importance," 56.
14. "These are questions we regard as insoluble and outside the domain of the positive philosophy; we, therefore, rightly abandon them to the imagination of the theologians or the subtleties of the metaphysicians." Comte praised the French mathematician and physicist Joseph Fourier, for instance, who studied the laws of heat but "not once inquired into the intimate nature of heat itself." Comte, "Nature and Importance," 57.
15. Comte, *Positive Philosophy*, 2:98.
16. But Newton's law of gravitation figured much differently in Comte's plan for a science of humans in motion than in Saint-Simon's. While Saint-Simon tried to use it as a fundamental principle of knowledge that spanned the natural and social sciences, Comte saw it as merely an exemplary case illustrating his fundamental principle that all sciences—those that address humans in motion as well as those that address matter in motion—resemble each other only in that they mature in similar ways, culminating in the positive stage.
17. The first volume, on mathematics, was published in 1830. Volumes 2 and 3, on astronomy and physics, and on chemistry and biology, respectively, were published in 1835. Volume 4, on the early, "dogmatic" part of the development of social physics, appeared in 1839; volume 5, on the history of social physics, with a discussion of the theological and metaphysical stages, appeared in 1841. Volume 6, Comte's principal discussion of social physics in the age of positivism, appeared in 1842.
18. Pickering, *Comte*, 1:486.
19. "Dogmatism is the normal state of human intelligence," Comte wrote, "toward which it tends by nature continually and in all genres even when it seems to be moving away from it the most." Cited in Pickering, *Comte*, 3:607.
20. "Rush Limbaugh on Energy & Oil," *On the Issues*, January 24, 2001, http://www.ontheissues.org/Celeb/Rush_Limbaugh_Energy_+_Oil.htm. The remark was made on Nov. 3, 2015.
21. Edmund DeMarche, "Michigan Congressman Who Believes in Climate Changes Says 'God Will Take Care of It,' " *Fox News Politics*, June 1, 2017, http://www.foxnews.com/politics/2017/06/01/michigan-congressman-who-believes-in-climate-change-says-god-will-take-care-it.html.
22. Sarah Palin, "Copenhgen=arrogance of man2think we can change nature's ways.MUST b good stewards of God's earth,but arrogant&naive2say man overpwers nature." 11:44 p.m., Dec. 18, 2009.
23. Mill to Comte, March 22, 1842, in *The Correspondence of John Stuart Mill and Auguste Comte*, tr. O. A. Haac (London: Transaction, 1995), 62.
24. Mill to Comte, June 15, 1843, in *The Correspondence of John Stuart Mill and Auguste Comte*, 164.
25. Comte to Mill, July 22, 1844, in *The Correspondence of John Stuart Mill and Auguste Comte*, 245.
26. Mill was distressed enough by Comte's views on women that, a few years after Comte's death, he wrote a book championing the equality of women, inspired by his frustrating exchanges. J. S. Mill, *The Subjection of Women* (London: Longmans, Green, 1869).
27. Pickering, *Comte*, 2:141.

28. Pickering, *Comte*, 2:143.
29. Pickering, *Comte*, 2:134.
30. Comte to Mill, December 18, 1845, in *The Correspondence of John Stuart Mill and Auguste Comte*, 339.
31. Comte to Mill, January 23, 1846, in *The Correspondence of John Stuart Mill and Auguste Comte*, 354.
32. Mill, *Auguste Comte and Positivism* (London: N. Trübner, 1865).
33. Pickering, *Comte*, 2:110.
34. In philosophical vocabulary, Mill viewed science as about epistemology, or what we could know about the world, while Comte thought that science sprang from our experiential encounter with the world.
35. Comte to Lewes, April 12, 1848, in Pickering, *Comte*, 2:102.
36. Pickering, *Comte*, 2:286.
37. In the Constituent Assembly, the Revolution's first legislature, the conservative factions sat to the right and the radical ones, such as the Jacobins, to the left of the president's chair, whence our modern terms "right wing" and "left wing."
38. One political historian has called the Positivist Society a "bizarre amalgamation of scientistic elitism and democratic populism." Pickering, *Comte*, 2:301.
39. Pickering, *Comte*, 2:8.
40. Pickering, *Comte*, 2:344.
41. Lewis A. Coser, *Masters of Sociological Thought*, 2nd ed. (Long Grove, IL: Waveland Press, 1977), 38.

CHAPTER SEVEN: MAX WEBER

1. John H. Marburger III, "Science's Uncertain Authority in Policy," in *Science Policy Up Close*, ed. R. P. Crease (Cambridge, MA: Harvard University Press, 2015), 229–37; R. P. Crease, "Why Don't They Listen?," *Physics World* (May 2014): 19. For the GAO report on the episode, see https://www.gao.gov/products/GAO-08–938R.
2. A valuable resource site for Weber can be found at http://sociosite.net/sociologists/weber_max.php. Last accessed 1 May 2018.
3. Marianne Weber, *Max Weber: A Biography*, tr. H. Zohn (London: Transaction, 1995), 77.
4. Marianne Weber, *Max Weber*, 202.
5. Max Weber, *From Max Weber: Essays in Sociology*, tr. H. H. Gerth and C. W. Mills (New York: Oxford University Press, 1946), 26.
6. Coser, *Masters*, xv.
7. In developing the concept of ideal types, which is one of Weber's foremost contributions to sociology, he took elements from both historians and scientists. From historians, he borrowed an appreciation of the need to look at the concrete reality in all its uniqueness, while from scientists he borrowed the reliance on abstractions and generalizations. Only such an "interpretive understanding" (*Verstehen*), as he called it, can allow the sociologist to study a value driven activity in a value-neutral way, and thereby clarify the ever-flowing "chaotic stream of events."
8. http://anthropos-lab.net/wp/wp-content/uploads/2011/12/Weber-objectivity-in-the-social-sciences.pdf.
9. Cited in Karl Jaspers, *On Max Weber*, ed. J. Dreijmanis, tr. R. Whelan (New York: Paragon, 1989), xvi.
10. Marianne Weber, *Max Weber*, 282.
11. Marianne Weber, *Max Weber*, 287.
12. Marianne Weber, *Max Weber*, 293.

13. Rationalization is not a value—it's a little wacky to say "I want to be rational"—and rationalization takes many forms. It can be practical, in the systematic pursuit of ends; substantive, in the determined attempt to realize a value; theoretical, in the creation of abstract concepts to control and command; or formal, in the imposition of rules and procedures for the sake of time, efficiency, and order—this being a distinctive feature of modern Western civilization.
14. Max Weber, *The Protestant Ethic and the Spirit of Capitalism*, tr. Talcott Parsons (New York: Routledge, 1992), 124.
15. Marianne Weber, *Max Weber*, 416.
16. Weber, *From Max Weber*, 217.
17. Weber, *From Max Weber*, 217.
18. Joachim Radkau, *Max Weber: A Biography*, tr. P. Camiller (Cambridge: Polity, 2011), 487.
19. Radkau, *Max Weber*, 487.
20. Schiller had referred to the "de-divinizing of the world" in "The Gods of Greece" in 1788; Weber used the phrase in "Science as a Vocation" (1917–1919). *From Max Weber*, 155.
21. *From Max Weber*, 155.
22. *From Max Weber*, 42.
23. Marburger, *Science Policy Up Close*, 229–37.

CHAPTER EIGHT: KEMAL ATATÜRK

1. Quoted in Niyazi Berkes, *The Development of Secularism in Turkey* (New York: Routledge, 1998), 79.
2. M. Alper Yalçinkaya, *Learned Patriots: Debating Science, State, and Society in the Nineteenth-Century Ottoman Empire* (Chicago: University of Chicago Press, 2015), 86.
3. Yalçinkaya, *Learned Patriots*, 13.
4. Yalçinkaya, *Learned Patriots*, 13.
5. For proponents of the new science, Yalçinkaya writes, "The old world and its representatives are associated with lethargy, esotericism, elitism, and unawareness, while the representatives of the new—that is, the holders of new knowledge—are portrayed as hardworking, truly beneficial subjects of the sultan, and true enlighteners of the people. Possessing new knowledge almost necessarily transforms one into a new kind of subject—one who is productive, industrious, and, ultimately, good." Yalçinkaya, *Learned Patriots*, 27.
6. Yalçinkaya, *Learned Patriots*, 57.
7. Wootton, *Invention of Science*, 107.
8. Yalçinkaya, *Learned Patriots*, 150.
9. Yalçinkaya, *Learned Patriots*, 101.
10. The fop, Yalçinkaya notes, sprang from the particular context of the Tanzimat period. "The discourse praising science emerged at the same time as the invasion of the Ottoman market by European consumer goods, the signing of treaties that guaranteed equal rights to non-Muslim subjects and the rise of non-Muslim Ottoman and European merchants who took advantage of the opening of Ottoman markets at the expense of Muslim subjects. The fop represented young Muslim men who, within such a context, wished to acquaint themselves with Europeans and live like them. These were men who tended also to hold a public post thanks to family connections and/or some education in the new schools of the empire. That statements about the benefits and significance of science were made by this particular group unavoidably shaped the way alternative discourses were constructed." Yalçinkaya, *Learned Patriots*, 101.

11. Yalçinkaya, *Learned Patriots*, 101.
12. Yalçinkaya, *Learned Patriots*, 98.
13. Yalçinkaya, *Learned Patriots*, 219–20.
14. M. Şükrü Hanioğlu, *Atatürk: An Intellectual Biography* (Princeton, NJ: Princeton University Press, 2011), 56.
15. Quoted in Berkes, *Development of Secularism*, 363.
16. Robert P. Crease, "Presidential Pledges," *Physics World* (January 2012): 19.
17. Ben Schreckinger, "Trump Acknowledges Climate Change—At His Golf Course," *Politico*, May 23, 1916, accessed May 1, 2018, https://www.politico.com/story/2016/05/donald-trump-climate-change-golf-course-223436.

CHAPTER NINE: EDMUND HUSSERL

1. Among other things, this work helped pave the way for general relativity.
2. William Finnegan, *Barbarian Days* (New York: Corsair, 2015), 293.
3. Paul Forman, "Weimar Culture, Causality, and Quantum Theory, 1918–1927: Adaptation by German Physicists and Mathematicians to a Hostile Intellectual Environment," in C. Carson, A. Kojevnikov, and H. Trischler, eds., *Weimar Culture and Quantum Mechanics: Selected Papers by Paul Forman and Contemporary Perspectives on the Forman Thesis* (Singapore: World Scientific, 2011), 90.
4. Cited in Dermot Moran, *Edmund Husserl: Founder of Phenomenology* (Malden, MA: Polity, 2005), 40.
5. David Carr, "Translator's Introduction" to *The Crisis of European Sciences and Transcendental Phenomenology* (Evanston, IL: Northwestern University Press, 1970), xix.

CHAPTER TEN: HANNAH ARENDT

1. Immanuel Kant, *The Conflict of the Faculties*, trans. M. Gregor (Lincoln: University of Nebraska Press, 1992), 95.
2. Hannah Arendt, " 'What Remains? The Language Remains': A Conversation with Günter Gaus," in *Essays in Understanding: 1930–1954*, ed. J. Kohn (New York: Harcourt, Brace, 1994), 6.
3. Hannah Arendt, "Martin Heidegger at Eighty," *New York Review of Books*, October 21, 1971, p. 51.
4. Arendt, "Martin Heidegger," p. 51.
5. Arendt, "Martin Heidegger," p. 51.
6. Hans Jonas, "Hannah Arendt: 1906–1975," in *Social Research* 43, no. 1 (Spring 1976): 3.
7. Elisabeth Young-Bruehl, *Hannah Arendt: For Love of the World* (New Haven, CT: Yale University Press, 1982), 49.
8. Young-Bruehl, *Hannah Arendt*, 51.
9. Young-Bruehl, *Hannah Arendt*, 301.
10. Arendt, *Essays*, 22.
11. Young-Bruehl, *Arendt*, 75.
12. Young-Bruehl, *Arendt*, 92.
13. Young-Bruehl, *Arendt*, 92.
14. Arendt, *Essays*, 12.
15. Arendt, *Essays*, 12.
16. Arendt, *Essays*, 14.
17. Arendt, *Essays*, 5.
18. Young-Bruehl, *Arendt*, 146.
19. Young-Bruehl, *Arendt*, 73.

20. Young-Bruehl, *Arendt*, 268.
21. http://www-leland.stanford.edu/dept/DLCL/files/pdf/hannah_arendt_we_refugees.pdf.
22. Hannah Arendt, "Organized Guilt and Universal Responsibility," in *Essays*, 121–32.
23. Young-Bruehl, *Arendt*, 257.
24. Quoted in Richard H. King, *Arendt and America* (Chicago: Universitiy of Chicago Press, 2015), 45.
25. Hannah Arendt, *The Origins of Totalitarianism* (New York: Harcourt, 1976), 297.
26. Hannah Arendt, *The Jewish Writings* (New York: Schocken, 2007), 48.
27. *Saturday Review*, March 24, 1951.
28. Arendt, *Essays*, p. 400.
29. Hannah Arendt, *The Human Condition* (Chicago: University of Chicago Press, 1958), 156.
30. King, *Arendt and America*, 75.
31. Arendt, *Human Condition*, 220.
32. Anne O'Byrne, *Natality and Finitude* (Bloomington: Indiana University Press, 2010), 82.
33. Arendt, *Human Condition*, 258.
34. Arendt, *Human Condition*, 251.
35. Arendt, *Human Condition*, 264.
36. Arendt, *Human Condition*, 284.
37. Arendt, *Human Condition*, 322.
38. Hannah Arendt, *Between Past and Future* (New York: Penguin, 2006), 105.
39. Arendt, *Between Past and Future*, 100.
40. Arendt, *Between Past and Future*, 109.
41. Arendt, *Between Past and Future*, 141.
42. I cannot discuss *On Revolution*, which compares the American and French revolutions, concluding that the reason for the enduring success of the former and failure of the latter was that the Americans had acquired "the habit of self-government" and sought to promote participation in the political sphere, while the French relied too much on theory and were motivated rather by the desire to achieve specific social goals such as the elimination of poverty. But Arendt warned against America's vulnerability to the culture of consumption and the politics of self-interest.
43. Hannah Arendt, *Crises of the Republic: Lying in Politics*; *Civil Disobedience*; *On Violence*; *Thoughts on Politics and Revolution* (New York: Mariner, 1972).
44. Arendt, *Crises*, 6.
45. Arendt, *Crises*, 6–7.
46. Arendt, *Crises*, 37.
47. Arendt, *Crises*, 42.
48. In 2010, Louisiana Governor Bobby Jindal invested $220 million in a scheme to use sand berms to block the 5 million barrel BP Gulf oil spill. Scientists dubbed the plan "Berm-doggle," saying it would harm local ecology and eventually wash away. When scientists at the National Oil Spill Commission said the berms were a failure and had captured only a "miniscule amount" of oil, Jindal attacked their report as "partisan revisionist history at taxpayer expense," likening his expert-defying actions to those of Huey Long, a populist predecessor.
49. Donald Trump, interview by Lesley Stahl, *60 Minutes*, CBS, October 15, 2018, https://www.cbsnews.com/news/donald-trump-full-interview-60-minutes-transcript-lesley-stahl-2018-10-14/.
50. Arendt, *Between Past and Future*, 55.
51. https://www.washingtonpost.com/news/energy-environment/wp/2018/02/07/

scott-pruitt-asks-if-global-warming-necessarily-is-a-bad-thing/?utm_term=.cf948b6201f2.

52. Robert P. Crease, "Tale of Two Anniversaries," *Physics World* (May 2007): 18.

53. George Monbiot, "Frightened by Donald Trump? You don't know the half of it," *The Guardian*, November 30, 2016, https://www.theguardian.com/commentisfree/2016/nov/30/donald-trump-george-monbiot-misinformation.

54. Arendt, *Between Past and Future*, 233–34.

55. Roger Pielke Jr., *The Honest Broker: Making Sense of Science in Policy and Politics* (Cambridge: Cambridge University Press, 2007), 4.

56. King, *Arendt in America*, 215.

CONCLUSION

1. Instances of science denialism seem to become more numerous every day. If one needs examples, Wikipedia's "Denialism" article is a place to start, last accessed May 1, 2018, https://en.wikipedia.org/wiki/Denialism.

2. A. A. Rosenberg et al., "Congress's attacks on science-based rules," *Science* 29 (May 2015): 964–66. Far stronger and more detailed denunciations have appeared since the start of the Trump presidency and are sure to continue.

INDEX

Page numbers in *italics* refer to illustrations. Page numbers after 286 refer to Notes.